# Springer Theses

Recognizing Outstanding Ph.D. Research

For further volumes:
http://www.springer.com/series/8790

# Aims and Scope

The series "Springer Theses" brings together a selection of the very best Ph.D. theses from around the world and across the physical sciences. Nominated and endorsed by two recognized specialists, each published volume has been selected for its scientific excellence and the high impact of its contents for the pertinent field of research. For greater accessibility to non-specialists, the published versions include an extended introduction, as well as a foreword by the student's supervisor explaining the special relevance of the work for the field. As a whole, the series will provide a valuable resource both for newcomers to the research fields described, and for other scientists seeking detailed background information on special questions. Finally, it provides an accredited documentation of the valuable contributions made by today's younger generation of scientists.

## Theses are accepted into the series by invited nomination only and must fulfill all of the following criteria

- They must be written in good English.
- The topic should fall within the confines of Chemistry, Physics, Earth Sciences and related interdisciplinary fields such as Materials, Nanoscience, Chemical Engineering, Complex Systems and Biophysics.
- The work reported in the thesis must represent a significant scientific advance.
- If the thesis includes previously published material, permission to reproduce this must be gained from the respective copyright holder.
- They must have been examined and passed during the 12 months prior to nomination.
- Each thesis should include a foreword by the supervisor outlining the significance of its content.
- The theses should have a clearly defined structure including an introduction accessible to scientists not expert in that particular field.

Adi Haber

# Metallocorroles for Attenuation of Atherosclerosis

Doctoral Thesis accepted by
the Technion—Israel Institute of Technology,
Haifa, Israel

*Author*
Dr. Adi Haber
Technion—Israel Institute of Technology
Haifa
Israel

*Supervisor*
Prof. Zeev Gross
Technion—Israel Institute of Technology
Haifa
Israel

*Co-supervisor*
Prof. Michael Aviram
Technion—Israel Institute of Technology
Haifa
Israel

ISSN 2190-5053
ISBN 978-3-642-30327-2
DOI 10.1007/978-3-642-30328-9
Springer Heidelberg New York Dordrecht London

ISSN 2190-5061 (electronic)
ISBN 978-3-642-30328-9 (eBook)

Library of Congress Control Number: 2012939121

© Springer-Verlag Berlin Heidelberg 2012
This work is subject to copyright. All rights are reserved by the Publisher, whether the whole or part of the material is concerned, specifically the rights of translation, reprinting, reuse of illustrations, recitation, broadcasting, reproduction on microfilms or in any other physical way, and transmission or information storage and retrieval, electronic adaptation, computer software, or by similar or dissimilar methodology now known or hereafter developed. Exempted from this legal reservation are brief excerpts in connection with reviews or scholarly analysis or material supplied specifically for the purpose of being entered and executed on a computer system, for exclusive use by the purchaser of the work. Duplication of this publication or parts thereof is permitted only under the provisions of the Copyright Law of the Publisher's location, in its current version, and permission for use must always be obtained from Springer. Permissions for use may be obtained through RightsLink at the Copyright Clearance Center. Violations are liable to prosecution under the respective Copyright Law.
The use of general descriptive names, registered names, trademarks, service marks, etc. in this publication does not imply, even in the absence of a specific statement, that such names are exempt from the relevant protective laws and regulations and therefore free for general use.
While the advice and information in this book are believed to be true and accurate at the date of publication, neither the authors nor the editors nor the publisher can accept any legal responsibility for any errors or omissions that may be made. The publisher makes no warranty, express or implied, with respect to the material contained herein.

Printed on acid-free paper

Springer is part of Springer Science+Business Media (www.springer.com)

**Parts of this thesis have been published in the following journal articles:**

Adi Haber, Michael Aviram, and Zeev Gross. **A Catalytic Antioxidant That Inhibits Cholesterol Biosynthesis**. *In preparation.*

Adi Haber, Michael Aviram, and Zeev Gross. **Variables That Influence Cellular Uptake and Cytotoxic/Cytoprotective Effects of Macrocyclic Iron Complexes**. *Inorg. Chem.* (2012), 51, 28–30.

Adi Haber, Michael Aviram, and Zeev Gross. **Protecting the beneficial functionality of lipoproteins by 1-Fe, a corrole-based catalytic antioxidant**. *Chem. Sci.* (2011) 2, 295–302.

Adi Haber, Atif Mahammed, Bianca Fuhrman, Nina Volkova, Raymond Coleman, Tony Hayek, Michael Aviram, and Zeev Gross. **Amphiphilic/Bipolar Metallocorroles That Catalyze the Decomposition of Reactive Oxygen and Nitrogen Species, Rescue Lipoproteins from Oxidative Damage, and Attenuate Atherosclerosis in Mice**. *Angew. Chem. Int. Ed.* (2008) 47, 7896–7900.

*Dedicated with love to my mother,
Anat Netzer*

# Supervisor's Foreword

Corroles have been the main research topic in my group ever since we discovered the one-pot synthesis of triarylcorroles in 1999. The immediate focus was on the synthesis of various derivatives and the application of their transition metal complexes for oxidation catalysis. Another synthetic discovery paved the way for the research described in this thesis: the extremely facile preparation of corroles with hydrophilic sulfonic acid head groups located on one pole of the otherwise lipophilic molecule. This allowed for the use of the metal complexes of the amphipolar corroles for biological and medicinal applications. In parallel, we realized that metallocorroles are much more efficient catalysts for reduction than for oxidation in both chemical and biomimetic systems. A particularly important finding was that they serve very well as catalysts for decomposition of reactive oxygen and nitrogen species (ROS and RNS, respectively) to biologically benign molecules/ions. Since ROS/RNS are involved in the development and progress of numerous diseases, we became interested in examining the feasibility of developing appropriate corrole metal complexes as therapeutic agents.

Adi, whose bachelor's degree was in Molecular Biochemistry, was the first graduate student of my (chemistry-only at that time) research group to be assigned such a project. The first step was to examine the interaction of the amphipolar corroles with various serum components, as the latter may be responsible for distribution of the former between the various body tissues and involved in cellular uptake. These investigations revealed very strong association of bis-sulfonated triarylcorroles to several serum proteins, but even stronger binding to lipoproteins. This important observation suggested that metal complexes of this corrole could display beneficial effects on preventing atherosclerosis, and consequently cardio-vascular diseases. Adi has examined the effects of three rationally designed corrole-metal complexes on the many variables that contribute to the development of atherosclerosis, with focus on both the "bad" (LDL) and the "good" (HDL) cholesterol carriers. The main novelty of Adi's thesis is that she has disclosed that the iron(III) complex of the bis-sulfonated corrole (**1-Fe**) prevents HDL from becoming dysfunctional due to oxidation and consequential loss of anti-athero-genic properties. **1-Fe** also reduced the pro-atherogenicity of the true variant of "bad cholesterol", oxidatively modified LDL. These and many other biochemical tests were encouraging enough for justifying the move toward animal-based models of atherosclerosis. Treatment of genetically engineered mice revealed that oral administration of **1-Fe** is highly beneficial for attenuating atherosclerosis development, both in young asymptomatic mice and in aged mice with already advanced signs of the disease.

The discoveries of this work were published in three international journals, cited in *Highlights in Chemical Science* 2011, and served as the bases for two patent applications. What is more, it paved the way for ongoing regulated preclinical studies that are needed for advancing metallocorroles as potential drugs for diseases where ROS/RNS have pronounced deleterious effects. All these achievements were made possible due to Adi's talent of managing to recruit for her research methods from the disciplines of chemistry, biochemistry, biology, and medicine.

Prof. Zeev Gross

# Acknowledgments

I would like to express my deepest gratitude to my supervisor, Prof. Zeev Gross, who gave me this great opportunity of exploring the activity of corroles in biological systems, a field that was new in his lab when I began my PhD studies. I would like to thank him for his trust and the great faith he had in me. Prof. Gross spent numerous hours listening/discussing/analyzing my results and consulting how to proceed with the research. Owing to his guidance I have learned how research should be conducted, and I am now an independent researcher who can face any new project. It has been a great pleasure working with Prof. Gross.

I would also like to thank my co-supervisor from the Rappaport medical school of the Technion, Prof. Michael Aviram. His invaluable advice and suggestions helped to forward the research toward a deeper understanding of the underlying mechanism of how corroles attenuate the development of atherosclerosis.

I would like to thank Prof. Bianca Fuhrman, who tutored me when I first started working on the biological aspects of my PhD research. Due to her excellent guidance I rapidly comprehended complex biochemical assays and how to implement these methods in my research.

I would further want to thank Aviva Lazarovich for listening, assisting, and encouraging whenever needed. Her good advice is highly appreciated.

I would also like to thank all the people who helped me out with the animal experiments: Dr. Atif Mahammed for synthesizing large quantities of the corroles; Prof. Tony Hayek for preparing blood and organs for analysis; Prof. Raymond Coleman for histological analysis of the atherosclerosis lesions; and Nina Volkova and Mira Rosenblat for helping me with animal handling and examination.

I would like to thank my family that was always there for me for moral and every other support needed. Last but not least I would like to thank my husband for being a great husband and a wonderful dad to our two special children.

I dedicate this thesis to my mother. There are not enough words to describe the help she gives me.

# Contents

1 **Introduction** .................................................... 1
   1.1  Biologically Relevant ROS/RNS ........................ 1
   1.2  Atherosclerosis ........................................ 3
        1.2.1  The Pro-Atherogenic Role of LDL and Macrophages ... 4
        1.2.2  The Anti-Atheroginicity of HDL ................... 6
        1.2.3  Common Treatments for Atherosclerosis ............. 6
   1.3  Corroles ................................................ 9
        1.3.1  Catalytic Antioxidants ............................ 10
        1.3.2  Electronic Spectra ................................ 11
   References ................................................... 12

2 **Research Goals** ................................................ 15

3 **Results** ........................................................ 17
   3.1  Interactions of Corroles with Lipoproteins ................. 17
        3.1.1  Corrole-Lipoproteins Binding ...................... 17
        3.1.2  Corrole Distribution in Serum ..................... 17
        3.1.3  Spectral Changes in the Presence
               of Lipoprotein Components ....................... 22
   3.2  The Effect of Corroles on Oxidative-Stress-Induced
       Lipoprotein Damage ...................................... 25
        3.2.1  Copper-Ions-Induced Oxidation .................... 25
        3.2.2  Peroxynitrite-Induced Oxidation ................... 27
        3.2.3  Lipoprotein Function .............................. 30
   3.3  Interactions of Corroles with Macrophages .................. 32
        3.3.1  Fluorescence-Based Detection Of 1-Ga ............. 32
        3.3.2  Chemiluminescence-Based Detection of 1-Fe ........ 34
        3.3.3  Cellular Accumulation of 1-Fe Versus
               Iron(III) Porphyrins ............................... 38

|  |  |  |  |
|---|---|---|---|
| | 3.4 | The Effect of Corroles on Macrophage Atherogenicity......... | 40 |
| | | 3.4.1 Effect on Oxidative-Stress-Induced Cell Death......... | 40 |
| | | 3.4.2 Effect on Macrophages Cholesterol Metabolism........ | 41 |
| | 3.5 | The Effect of Corroles on the Development of Atherosclerosis..................................... | 44 |
| **4** | **Discussion**................................................. | | 45 |
| | 4.1 | Interaction of Corroles with Lipoproteins.................. | 45 |
| | | 4.1.1 Corrole Distribution in Serum.................... | 45 |
| | | 4.1.2 Mode of 1-Fe Binding to Lipoproteins.............. | 47 |
| | 4.2 | The Effect of Corroles on Oxidative-Stress Induced Lipoprotein Damage.................................. | 50 |
| | | 4.2.1 Lipoprotein Oxidation Products.................... | 50 |
| | | 4.2.2 Lipoproteins Biological Activities.................. | 52 |
| | | 4.2.3 Mechanistic Analysis............................ | 53 |
| | 4.3 | Interactions of Corroles with Macrophages................. | 54 |
| | | 4.3.1 Chemiluminescence-Based Detection of 1-Fe Accumulation in Macrophages.............. | 55 |
| | | 4.3.2 Comparison of Accumulation in Macrophages of 1-Fe and Related Iron(III) Porphyrins............. | 57 |
| | 4.4 | The Effect of Corroles on Macrophage Atherogenicity........ | 58 |
| | | 4.4.1 Oxidative-Stress Induced Damage.................. | 58 |
| | | 4.4.2 Effect on Cholesterol Metabolism.................. | 60 |
| | 4.5 | The Effect of Corroles on Atherosclerosis Development in Mice................................... | 61 |
| | References................................................. | | 62 |
| **5** | **Conclusions**................................................ | | 67 |
| **6** | **Materials and Methods**...................................... | | 71 |
| | 6.1 | General Methods...................................... | 71 |
| | | 6.1.1 Materials...................................... | 71 |
| | | 6.1.2 Measurement of Lipoprotein Oxidation Products....... | 72 |
| | | 6.1.3 Cell Cultures................................... | 74 |
| | | 6.1.4 Cellular Assays................................. | 74 |
| | | 6.1.5 Histopathological Development of Atherosclerotic Lesions in Mice................................. | 76 |
| | 6.2 | Experimental Procedures................................ | 76 |
| | | 6.2.1 Corrole-Lipoproteins Binding...................... | 76 |
| | | 6.2.2 Corrole Distribution in Serum..................... | 77 |
| | | 6.2.3 Spectral Changes in the Presence of Lipoprotein Components.................................... | 77 |
| | | 6.2.4 Lipoprotein Oxidation........................... | 79 |
| | | 6.2.5 Lipoprotein Functions........................... | 80 |

|   |   |   | |
|---|---|---|---|
|   | 6.2.6 | Interactions of Corrole and Porphyrin Complexes with Cells | 81 |
|   | 6.2.7 | Effect of Corrole and Porphyrin Complexes on Cells | 83 |
|   | 6.2.8 | The Effect of Corroles on the Development of Atherosclerosis | 85 |
| References | | | 86 |

**Appendix** ............................................. 87

# Symbols and Abbreviations

| | |
|---|---|
| **1-Fe** | Fe(III) 5,10,15-tris(pentafluorophenyl)-2,17 bis(sulfonic acid) corrole |
| **1-Ga** | Ga(III) 5,10,15-tris(pentafluorophenyl)-2,17 bis(sulfonic acid) corrole |
| **1-Mn** | Mn(III) 5,10,15-tris(pentafluorophenyl)-2,17 bis(sulfonic acid) corrole |
| **2-Fe** | Fe(III) meso-Tetra(4-sulfonatophenyl)porphine chloride |
| **3-Fe** | Fe(III) meso-Tetra(N-Methyl-4-Pyridyl)porphine pentachloride |
| ACAT | Acetyl:Coenzyme A acetyltransferase |
| apoAI | Apolipoprotein AI |
| apoAI-KO mice | ApoAI knock out mice |
| apoB$_{100}$ | Apolipoprotein B100 |
| apoE | Apolipoprotein E |
| BSA | Bovine serum albumin |
| CD | Circular dichroism |
| cDNA | Complementary DNA |
| CETP | Cholesterol ester transfer protein |
| CVD | Cardiovascular diseases |
| DAPI | 4′,6-Diamidino-2-Phenylindole |
| DCF | 2,7-Dichlorofluorescein |
| DCFH | 2,7-Dichlorodihydrofluorescein |
| DCFH-DA | DCFH Diacetate |
| DiOC6 | 3,3′-Dihexyloxacarbocyanine Iodide |
| DMEM | Dulbecco's modified eagle medium |
| DNA | Deoxyribonucleic acid |
| DTPA | Diethylene triamine pentaacetic acid |
| E$^0$ mice | ApoE deficient mice |
| EDTA | Ethylene diamine tetraacetic acid |
| ELISA | Enzyme-linked immunosorbent assay |
| EtCys | Ethyl cysteine ester |
| FACS | Fluorescence-activated cell sorting |

| | |
|---|---|
| FCS | Fetal calf serum |
| FITC | Fluorescein isothiocyanate |
| FS | Fluvastatin |
| GAPDH | Glyceraldehyde 3-phosphate dehydrogenase |
| HDL | High density lipoproteins |
| HDL2 | HDL subfraction 2 |
| HDL3 | HDL subfraction 3 |
| HDL-C | HDL cholesterol |
| His | Histidine |
| HMG-CoA | 3-Hydroxy-3-methyl-glutaryl-CoA |
| HMGCR | HMG reductase |
| HPLC | High pressure liquid chromatography |
| HRP | Horseradish peroxidase |
| HSA | Human serum albumin |
| LCAT | Lecithin-cholesterol acyltransferase |
| LC-MS | Liquid chromatography—mass spectrometry |
| LDL | Low density lipoproteins |
| LDL-C | LDL cholesterol |
| LPDS | Lipoprotein deficient serum |
| MeArg | Methyl arginine ester |
| MeHis | Methyl histidine ester |
| MeLys | Methyl lysine ester |
| MeMet | Methyl methionine ester |
| MeTyr | Methyl tyrosine ester |
| MFI | Mean fluorescence intensity |
| MPM | Mouse peritoneal macrophages |
| mRNA | Messenger RNA |
| MTT | 3-(4,5-Dimethylthiazol-2-yl)-2,5-diphenyltetrazolium bromide |
| NADH | Nicotinamide adenine dinucleotide |
| NADPH | Nicotinamide adenine dinucleotide phosphate |
| NOS | Nitric oxide synthase |
| oxHDL | Oxidized HDL |
| oxLDL | Oxidized LDL |
| PAGE | Polyacrylamide gel electrophoresis |
| PBS | Phosphate buffered saline |
| PCR | Polymerase chain reaction |
| PON1 | Paraoxonase 1 |
| PON1-KO mice | PON1 Knock out mice |
| PS | Pravastatin |
| RNA | Ribonucleic acid |
| RNS | Reactive nitrogen species |
| ROS | Reactive oxygen species |
| RPMI | Roswell Park Memorial Institute medium |
| SDS | Sodium dodecyl sulfate |
| SIN-1 | 3-Morpholino sydnonimine |

| | |
|---|---|
| SOD | Superoxide dismutase |
| TBA | Thiobarbituric acid |
| TLC | Thin layer chromatography |
| TMB | 3,3′,5,5′-Tetramethylbenzidine |
| VLDL | Very low density lipoproteins |

# Chapter 1
# Introduction

## 1.1 Biologically Relevant ROS/RNS

"Free radicals" is the term commonly used for molecules or ions that contain an odd number of electrons. The unavoidable presence of (at least) one unpaired electron has an enormous impact on the chemical reactivity of free radicals. They react very fast with non-radical species by either abstraction of an electron (acting as an oxidizing agent), donation of an electron (acting as a reducing agent), or by attachment to the non-radical [1]. The product formed in the latter case (commonly termed secondary radical) also contains an unpaired electron, and hence may react with another non-radical and propagate a chain reaction. An example of such a process is lipid peroxidation, discussed in details in Sect. 4.2.1 (p. 50).

The term reactive oxygen species (ROS) includes not only oxygen-centered free radicals, but also neutral molecules that are either oxidizing agents or easily converted into radicals. Oxygen-based radicals include superoxide ($O_2^-$), peroxyl radical ($RO_2\cdot$), alkoxyl radical ($RO\cdot$), and hydroxyl radical ($\cdot OH$), while hydrogen peroxide ($H_2O_2$), hypochlorous acid (HOCl), and ozone ($O_3$) are non-radical ROS. Similarly, reactive nitrogen species (RNS) include the radicals nitric oxide ($\cdot NO$) and nitric dioxide ($\cdot NO_2$) and the non-radical peroxynitrite ($ONOO^-$).

ROS and RNS are formed within the body due to exposure to radiation and pollution, and also during normal biological processes. Minute levels of these molecules fulfill some very important physiological roles, e.g., the radical $\cdot NO$ is both an important signal transduction molecule and a powerful vasodilator. Elevation in the amount of ROS/RNS leads to oxidative modifications of lipids, proteins, sugars, DNA and other vital biomolecules. This eventually results in serious biological malfunctions and, accordingly, the levels of ROS/RNS need to be strictly controlled. Natural molecules with antioxidant properties and specific enzymes that either neutralize the primary oxidizing species or interfere with the radical chain reactions fulfill this mission [2]. Depletion of antioxidants and/or overproduction of ROS/RNS leads to a situation called oxidative stress, which is involved in the development and progression of numerous diseases. Oxidative

**Fig. 1.1** The main ROS/RNS in the body. The mild oxidant superoxide may transform to the more toxic hydrogen peroxide or, in the presence of nitric oxide, to peroxynitrite. The two latter species lead to the formation of the very strong oxidants, hydroxyl radical and nitrogen dioxide radicals

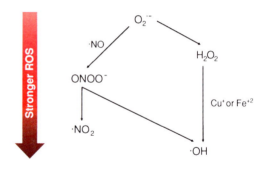

**Scheme 1.1** The action of ROS decomposition enzymes. The reactions catalyzed by: SOD (**a**), catalase (**b**) and glutathione peroxidase (**c**)

(a) $2O_2^{\bullet-} + 2H^+ \xrightarrow{SOD} H_2O_2 + O_2$

(b) $H_2O_2 \xrightarrow{Catalase} H_2O + \frac{1}{2}O_2$

(c) $H_2O_2 + 2GSH \xrightarrow{Glutathione\ Peroxidase} H_2O + 2GSSG$

stress increases with age due to decreased activity of antioxidant enzymes and is hence implicated in aging and in many neurodegenerative diseases typical for older individuals [3].

The main ROS/RNS formed in the body are depicted in Fig. 1.1. One source for ROS production within all cells is the respiratory chain, which is responsible for the four-electron reduction of $O_2$ to water. Uncoupling of the delicate processes occurring in mitochondria causes leakage of electrons from the electron carriers, and results in one-electron reduction of oxygen to $O_2^-$ [4]. While this is accidental, much of the $O_2^-$ formed in the body is functional. For instance, activated phagocytes produce this radical for killing the bacteria that they engulf [5]. This is achieved by NAD(P)H oxidases, which utilize nicotinamide adenine dinucleotide (NADH) or the nicotinamide adenine dinucleotide phosphate (NADPH) as electron donors for one-electron reduction of molecular oxygen. Vascular smooth muscle cells and endothelial cells also contain NAD(P)H oxidases, and these enzymes are a major source for ROS production in the vasculature [6].

$O_2^-$ is a relatively weak oxidant, which is effectively eliminated by the superfast acting family of enzymes called superoxide dismutases (SOD, Scheme 1.1) [7]. Note, however, that while this dismutation reaction eliminates a free radical it leads to the production of a very harmful free radical precursor, $H_2O_2$. The cytotoxicity of $H_2O_2$, which may also be formed by the action of oxidases on molecular oxygen, stems from its Fenton reaction with transition metal ions. The outcome is formation of the extremely reactive and damaging hydroxyl radical (·OH). According to the above, complete neutralization of $O_2^-$ requires the coupling of SOD with enzymes that either use or eliminate $H_2O_2$. Enzymes that perform the latter function are catalases, which decomposes $H_2O_2$ to molecular oxygen and water, and glutathione peroxidase, whose function relies on oxidation of reduced glutathione (GSH) by $H_2O_2$ (Scheme 1.1) [8]. It is also important to realize that Fenton chemistry is catalyzed by improperly chelated transition metal ions (frequently called "free metals"), but not by naturally occurring metal-containing proteins, such as the iron-transporting transferrin and heme-based proteins/enzymes.

An alternative fate for $O_2^-$ is reaction with ·NO, formed by nitric oxide synthases (NOS) that catalyze the oxidation of L-arginine to L-citrulline [9]. Enzymes of the NOS family are expressed in various cells in the cardiovascular system, and it has been shown that in some cases the enzyme may get uncoupled and produce not only ·NO but also $O_2^-$ [10]. The reaction between the two radicals, which is even faster than the decomposition of $O_2^-$ by SOD, creates a new toxin: $ONOO^-$ [11]. Most of the anion is protonated to its acidic form under physiological pH (pKa = 6.8), and the thus formed peroxynitrous acid (ONOOH) is partial converted to the highly toxic ·OH and ·$NO_2$ radicals [12]. There is no known enzyme for the detoxification of ONOOH, and the reactions of the main natural antioxidant molecules with this species are relatively inefficient (with rate constants of $1.5 \cdot 10^3$ $M^{-1}$ $s^{-1}$ for glutathione, the major cellular antioxidant, and $2.3 \cdot 10^2$ $M^{-1}$ $s^{-1}$ for ascorbate, the major antioxidant in plasma) [13].

## 1.2 Atherosclerosis

Atherosclerosis is a chronic vascular disease, in which the arteries undergo thickening and lose their elasticity as a result of cholesterol sedimentation in the artery wall (Fig. 1.2). In the early stages of the disease cholesterol accumulates within arterial macrophages, transforming them to lipid-loaded foam cells. Extensive atherosclerosis narrows the artery lumen thus reducing blood flow, and may develop to complete blockage of the artery. Three arteries are the major sites of atherosclerosis: the coronary arteries, the cerebral arteries and the aorta. Complete blood flow obstruction results in myocardial infarction (heart attack) or stroke.

The pathology of the disease involves interaction between artery wall cells, blood cells and plasma lipoproteins [14, 15]. Lipoproteins are the vehicles that carry lipids in the bloodstream (additional data on lipoproteins may be found in Table 4.1, p. 50). Very low density lipoprotein (VLDL) particles are synthesized by the liver and their function is to transport fatty acids to adipose tissue and muscle. After triglyceride removal in peripheral tissues, a portion of the remaining VLDL is metabolized to low density lipoprotein (LDL) particles by further removal of core triglycerides and dissociation of apolipoproteins. In humans, the majority of serum cholesterol is carried by LDL particles, which are responsible for delivery of lipids to peripheral tissues. LDL particles are cell-internalized by endocytosis following binding to LDL receptors, and when they arrive to lysosomes the cholesterol esters of LDL are hydrolysed and released to the cell

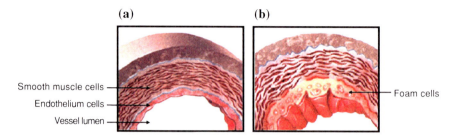

**Fig. 1.2** Normal and atherosclerotic arteries. Illustrations of a normal artery (**a**) and of an atherosclerotic artery, in which foam cells have developed in the vessel wall (**b**)

(while the LDL receptors are recycled back to the plasma membrane). Within the cell free cholesterol may undergo esterification by acyl CoA:cholesterol acyltransferas (ACAT) for storage in lipid droplets. These cholesterol esters can later be hydrolyzed by hormone-sensitive lipase, generating free cholesterol for incorporation into membranes. Several processes regulate normal cholesterol levels, which may be divided into uptake and removal events. Excess of intracellular cholesterol inhibits cholesterol biosynthesis within the cell (further described in Sect. 1.2.3.1, p. 7), as well as the expression of LDL receptor on the cell membrane [16]. Removal of cholesterol is performed by high density lipoproteins (HDL) in a process called reverse cholesterol transport, further described in Sect. 1.2.2 (p. 6).

## 1.2.1 The Pro-Atherogenic Role of LDL and Macrophages

Circulating LDL is relatively protected against oxidation owing to the high concentration of antioxidants in the blood. However, the endothelium and smooth muscle cells of the artery wall produce radical species as part of their normal function (described in Sect. 1.1, p. 1), thus creating high local concentrations of ROS/RNS and leading to oxidation of LDL in the sub-endothelium space [17]. Elevation in plasma LDL levels leads to a higher steady state concentration in the vessel wall. High blood LDL levels are indeed considered a major risk factor of cardiovascular diseases (CVD). During oxidation, polyunsaturated fatty acids in the cholesteryl esters, phospholipids and triglycerides present in LDL undergo radical chain reactions to yield a broad array of fragments. The sole protein present in LDL, apolipoprotein $B_{100}$ (apo$B_{100}$), may also be fragmented or modified by covalent linkage of the lipid fragments to lysine residues.

Normal arteries do not contain oxidized LDL (oxLDL) [18], but lipoproteins fractionated from atherosclerotic lesions are highly enriched with oxLDL, approximately 70 times higher than the level in plasma LDL of the same patient [19]. The mean concentration of oxLDL in plasma is also elevated in patients with

## 1.2 Atherosclerosis

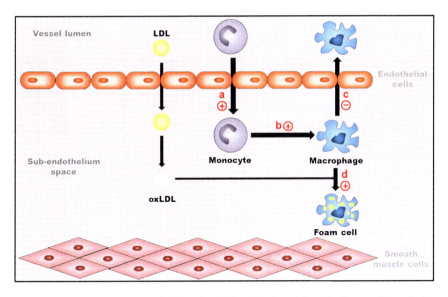

**Fig. 1.3** The pro-atherogenic activities of oxLDL. LDL is oxidized in the arterial wall. The resultant oxLDL recruits monocytes to the artery wall (*a*), promotes their differentiation into macrophages (*b*), reduces macrophages escape from the artery wall (*c*), and supports transformation of macrophages to foam cells (*d*)

CVD relative to normal subjects, 1.5-fold for patients with carotid atherosclerosis [19], and 3.4-fold for patients with myocardial infarction [20]. Plasma LDL from patients with high risk to develop atherosclerosis, such as hypercholesterolemic and diabetic, also demonstrate increased susceptibility to ex vivo oxidation compared with LDL from normal subjects [21]. Nitrotyrosine-modified LDL, a specific marker for the presence of RNS, is up to 80-fold elevated in atherosclerotic lesions relative to circulating LDL [22].

Several atherogenic properties are displayed by oxLDL (Fig. 1.3): they cause the recruitment of monocytes (circulating leukocytes) to the artery wall, promote the differentiation of these monocytes into macrophages, inhibit the ability of the macrophages to leave the artery wall, support the transformation of macrophages to foam cells, and more [17]. The recruited macrophages release large amount of radicals, consequently leading to increased local oxidative stress and to subsequent LDL oxidation. While native LDL is taken up by macrophages in a regulated manner by the LDL receptor, the internalization of oxLDL is mediated by scavenger receptors that are not controlled by the presence of excess cholesterol. This causes the accumulation of cholesterol within the macrophages and thus to their transformation to foam cells, the first step in the development of an atherosclerotic plaque. So while the recruitment of monocytes and their differentiation into macrophages may initially serve a protective function by removing cytotoxic oxLDL particles, progressive accumulation of macrophages ultimately leads to development of atherosclerotic lesions.

## 1.2.2 The Anti-Atheroginicity of HDL

HDL exhibits several anti-atherogenic properties, and accordingly low plasma levels of HDL are considered a risk factor of CVD. Reverse cholesterol transport is considered a major cardioprotective property of HDL, since it commences with the efflux of free cholesterol from cell membranes to lipid-poor nascent HDL. This process is largely dependent on apolipoprotein A-I (apoAI), the most abundant apolipoprotein in HDL [23]. Subsequent esterification of cholesterol by lecithin-cholesterol acyltransferase (LCAT) generates small spherical HDL (subfraction 3, HDL3) and, in turn, large spherical HDL (subfraction 2, HDL2). These spherical HDL contain a lipid core of mainly cholesterol ester, which can then be exchanged for triglycerides with other lipoproteins by the action of cholesterol ester transfer protein (CETP) or be cleared from the body by specific uptake to the liver. Another important role of HDL is its antioxidant activity. HDL serves as a "sink" for lipid hydroperoxides, which it extracts from oxidized membranes and from oxLDL and transports to the liver for detoxification. ApoAI is a major component of this activity, as it is responsible for extraction of the oxidized lipids [24]. Several HDL-associated enzymes, mainly paraoxonase 1 (PON1), also contribute to the anti-oxidative activity of HDL by hydrolysis of oxidized phospholipids that accumulate in HDL [25].

It is important to realize that HDL is even more prone to oxidation than LDL because of its lower antioxidant content and its higher concentration in the artery wall. The latter is due to its smaller size and its greater serum concentration.[1] Plasma of diabetes and coronary artery disease patients show a 1.5-fold increase in oxidized HDL (oxHDL) concentrations indeed [26]. Nitrotyrosine-modified HDL levels are also increased in subjects with cardiovascular disease [27]. Oxidation of HDL alters both its lipid and its protein composition, and impairs its beneficial activities [28]. This includes reduced LCAT activity even under subtle oxidation, decreased PON1 activity in correlation with increased HDL oxidation, and more. The awareness to dysfunctional HDL is steadily increasing during the last decade.

## 1.2.3 Common Treatments for Atherosclerosis

Risk factors for development of atherosclerosis include obesity, hypertension, smoking, lack of exercise and more. Therapeutic means can be designed against many steps in the atherogenic process, including those affecting LDL levels and

---

[1] When conducting blood tests the value measured is the amount of cholesterol in LDL and HDL (LDL-C and HDL-C, respectively), and so higher values are related to the cholesterol rich LDL (typically two to threefold more than for HDL), although on a molar basis the concentration of HDL is much higher (see Table 4.1, p. 50).

1.2 Atherosclerosis

oxidation, reverse cholesterol transport, and others. Two main methods are statin treatment and antioxidant consumption.

### 1.2.3.1 Statins

The main dogma for prevention of CVD is to control plasma cholesterol levels by an appropriate diet, and if necessary by pharmaceuticals [29]. The leading drugs for lowering serum LDL are statins (Scheme 1.2), which have increased the lifespan of humans by up to 4 years, more than all other drugs developed during the last 30 years [30]. However, statin treatment still reduces cardiovascular events by only 30% [31], as this multi-factorial disease requires more than just balancing of serum cholesterol levels. Treatment with high doses of statins (80 mg) is recently turning out to be highly problematic due to severe side effects, including muscle pain and a negative impact on diabetes development [32, 33].

Statins act mainly by affecting the de novo biosynthesis of cholesterol. The biosynthetic pathway is a multistep process, with one of the earliest steps being the transformation of 3-hydroxy-3-methyl-glutaryl-coenzyme A (HMG-CoA, Scheme 1.2) to mevalonate. This committing step of the synthesis is catalyzed by the enzyme HMG-CoA reductase (HMGCR). The natural control of the enzyme is achieved by feedback inhibition, with cholesterol reducing HMGCR gene expression by reducing mRNA transcription and protein translation and increasing protein degradation. A second level of control, which balances the active pool of the enzyme, is deactivation of the protein caused by its phosphorylation. Statins act by competitive inhibition of the HMGCR enzyme, with the main target being the liver. Reduced cellular cholesterol biosynthesis increases cellular intake of cholesterol, thus removing LDL from the plasma and lowering its chance to become oxidized.

### 1.2.3.2 Dietary Antioxidants

Vitamin E ($\alpha$-tocopherol, Scheme 1.3) is the most important lipophilic antioxidant in biological systems. It is situated within lipoproteins and rapidly reacts with lipid peroxides, thus interfering with the propagation of lipid peroxidation, the first step in lipoprotein oxidation (for more details see Sect. 4.2.1, p. 50). This reaction yields a tocopheryl radical, which may then react with an additional lipid peroxide to terminate the radical reaction. However, at low radical fluxes the tocopheryl radical may react with a poly-unsaturated fatty acid before it encounters a second lipid peroxide, thus serving as a pro-oxidant [34]. Alternatively, $\alpha$-tocopherol may be regenerated from its radical by ascorbic acid (vitamin C, Scheme 1.3), the main water soluble antioxidant. As $\alpha$-tocopherol is present within lipoproteins in very low amount (only 8–12 molecules per each LDL particle and less than 1 molecule per each HDL particle) [28], this recycling by ascorbate is very important, and is responsible for the efficient protection of lipoproteins against oxidation within the

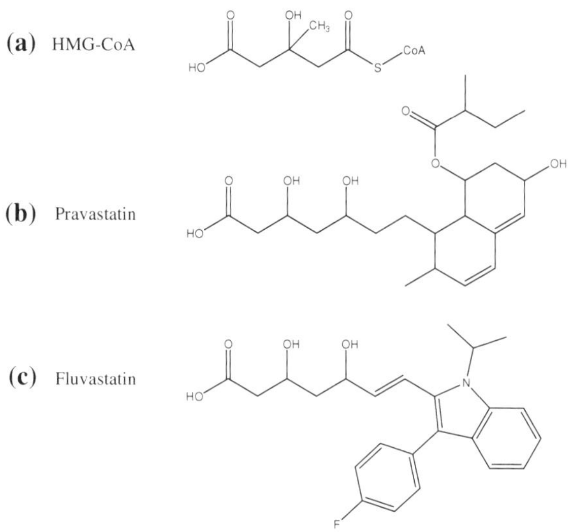

**Scheme 1.2** The chemical structure of HMG-CoA and selected statins. HMG-CoA (**a**), pravastatin (**b**) and fluvastatin (**c**)

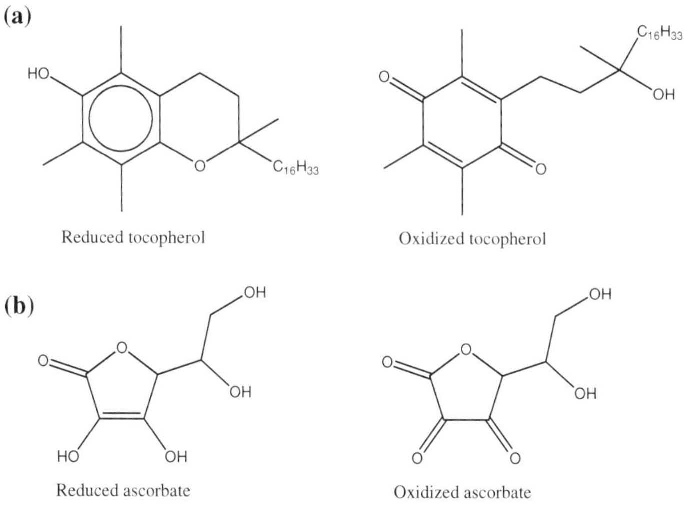

**Scheme 1.3** The chemical structure of main antioxidants in the body. Reduced and oxidized tocopherol (**a**), and reduced and oxidized ascorbate (**b**)

circulation. However, ascorbic acid may also act as a pro-oxidant as it reduces transition metal ions to a low valence state (Cu(I), Fe(II), etc.) in which they catalyze the transformation of lipidic hydroperoxides (or hydrogen peroxide) into radicals [34].

1.2 Atherosclerosis

Concentrations of α-tocopherol and ascorbic acid are mainly influenced by dietary intake. The involvement of oxidative stress in the development of atherosclerosis thus implicates an antioxidant rich diet as a means for reducing plaque formation. Indeed, consumption of food rich with these and other dietary antioxidants (such as pomegranate juice and red wine) have been shown to reduce the risk of atherosclerosis [21, 35], though vitamin supplements were highly disappointing [36, 37]. While bioavailability is one concern regarding the inefficiency of dietary antioxidants, the main problem is their sacrificial mode of action—one antioxidant molecule detoxifies only one (or maximum two) radical(s).

## 1.3 Corroles

Corroles are tetrapyrolic macrocycles of structural skeletal resemblance to the cobalt-chelating corrin molecule of vitamin $B_{12}$. In contrast with the corrin, the corrole molecule is fully unsaturated and of aromatic conjugation, and hence more closely related to porphyrins, the iron-chelating prosthetic groups present in numerous heme enzymes and proteins (e.g., hemoglobin and cytochromes). The basic skeleton of the corrole contains one carbon atom less than that of the porphyrin, making its core smaller than that of the porphyrin (scheme 1.4). In addition, while porphyrins are di-anionic ligands, the corroles are tri-anionic. Porphyrins have been extensively utilized for many years in various applications, however the field of corroles (first discovered at 1964) remained almost futile up to 1999, when a very simple and efficient methodology for the synthesis of triarylcorroles was discovered [38]. Later, water soluble derivatives of these corrole were developed [39, 40], allowing for their employment in biological applications [41–47]. The lower symmetry of the corrole relative to the porphyrin allowed for easy access to derivatives that have polar substituents on only one side of the corrole molecule, thus forming an amphipolar condition.

The *meso* positions (the carbon atoms bringing between the pyrol rings) of corroles and porphyrins are very reactive and easily oxidized, and hence *meso*-substituted derivatives of both macrocycles, such as those depicted in scheme 1.5, are more stable and suitable for various applications [48]. Despite of this ease of oxidation, corroles still stabilize high oxidation states of their chelated metal ions more effectively than most ligands. This is due to their small core together with the high anionic charge, thus turning them to strong $\sigma$ donors that considerably raise the energy of the metal d orbitals and reduce the metal's oxidation power [49]. In contrast with "free iron", and most other iron complexes, which mainly increase oxidative stress, iron corrole complexes are actually reducing agents, and they do not release the metal even under extreme conditions. These characteristics paved the way for using transition metal complexes of corroles for the detoxification of ROS/RNS, as described in Sect. 1.3.1 (p. 10).

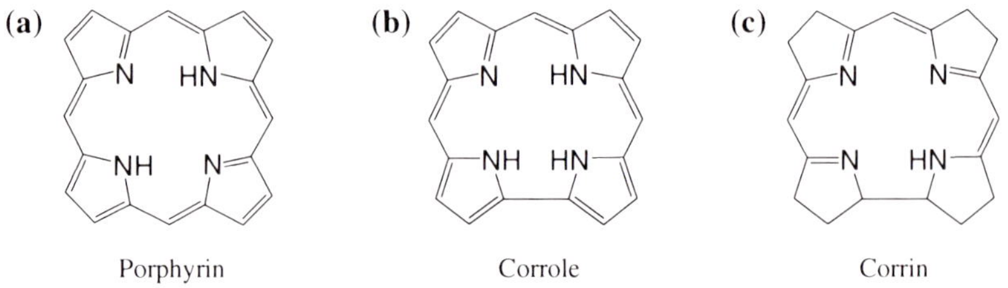

10                                                                                    1  Introduction

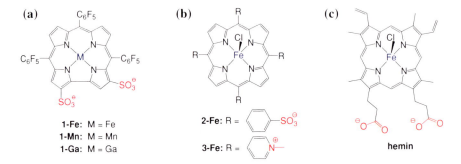

**Scheme 1.4** The basic skeleton of selected macrocyclic ligands. The basic skeleton of porphyrins (**a**), corroles (**b**) and corrin (**c**)

**Scheme 1.5** The chemical structure of the compound discussed in this research. Bis-sulfonated corrole complexes (**a**), synthetic iron-porphyrins (**b**) and the natural iron porphyrin (**c**)

The structures of the compounds relevant for this research, the bis-sulfonated corrole metal complexes (**1-Fe**, **1-Mn** and **1-Ga**) and the natural (hemin) and synthetic iron porphyrin complexes (**2-Fe** and **3-Fe**), are outlined in Scheme 1.5.

## 1.3.1 Catalytic Antioxidants

Dietary (sacrificial) antioxidants are rapidly and irreversibly consumed in the body, and thus display limited effectiveness against oxidative stress (Sect. 1.2.3.2, p. 7). The use of antioxidant enzymes as therapeutic agents to attenuate ROS-induced damage has also had mixed success, as due to their large size they display low cell permeability, short circulating half-life, and antigenicity [50]. One problem shared by both antioxidant groups is their non-efficiency against peroxynitrite, a reactive species that damages lipoproteins and the arterial wall. To overcome many of these limitations, an increasing number of low molecular-weight *synthetic catalytic antioxidants* have been developed, which serve as SOD and catalase mimetics, and may also decompose peroxynitrite in a catalytic manner [51, 52]. Transition metal complexes of corroles and porphyrins are such

1.3 Corroles

**Table 1.1** Catalytic rates for ROS/RNS decomposition by **1-Fe** and **1-Mn**

| Compound | $O_2^-$ decomposition $[M^{-1} s^{-1}]$ [53] | $H_2O_2$ decomposition $[M^{-1} s^{-1}]$ [54] | ONOOH decomposition $[M^{-1} s^{-1}]$ [55] |
|---|---|---|---|
| **1-Fe** | $3 \times 10^6$ | 6400 | $3 \times 10^6$ |
| **1-Mn** | $5 \times 10^5$ | None | $9 \times 10^4$ |

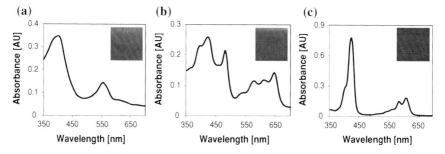

**Fig. 1.4** The absorption spectra of **1-Fe**, **1-Mn** and **1-Ga** in buffer. The UV–vis absorption spectra of buffered solutions of **1-Fe** (**a**), **1-Mn** (**b**) and **1-Ga** (**c**) and the color of each solution (insert)

catalytic antioxidants. The catalytic rates for decomposition of $O_2^-$, $H_2O_2$ and ONOOH by the corroles **1-Fe** and **1-Mn** are given in Table 1.1.

Several important features should be mentioned regarding the two corrole complexes. First, **1-Fe** is more efficient than **1-Mn** for the decomposition of all three main physiologically relevant ROS/RNS. Secondly, while **1-Fe** decomposes peroxynitrite by an isomerization mechanism to nitrite ($NO_3^-$), **1-Mn** acts by disproportionation to form nitrate ($NO_2^-$). Third, **1-Mn** is the first manganese complex to display true catalytic activity, without the need for supplementation with reducing agents [56]. Fourth, although the *most optimized* porphyrin-based catalytic antioxidants display higher catalytic rates regarding $O_2^-$ decomposition and similar rates regarding ONOOH detoxification, they are very poor catalase mimetics and their synthesis is a long and tedious task.

## 1.3.2 Electronic Spectra

Corroles display rich absorbance spectra at the visible range, which serve as an important tool for their detection and characterization, and will hence be extensively discussed throughout this text. The electronic spectra of corroles are typically composed of strong ($\varepsilon = 10^5$–$10^6$ $M^{-1} \cdot cm^{-1}$) near-UV absorption(s), termed Soret band(s), and of weaker ($\varepsilon = 10^3$–$10^4$ $M^{-1} \cdot cm^{-1}$) visible range absorptions, termed Q bands. The UV–vis absorbance spectra of buffered solutions of the corroles relevant for this research are depicted in Fig. 1.4.

# References

1. Slater, T.F.: Free-radical mechanisms in tissue injury. Biochem. J. **222**, 1–15 (1984)
2. Cadenas, E.: Basic mechanisms of antioxidant activity. BioFactors **6**, 391–397 (1997)
3. Finkel, T., Holbrook, N.J.: Oxidants, oxidative stress and the biology of ageing. Nature **408**, 239–247 (2000)
4. Fridovich, I.: Superoxide radical: an endogenous toxicant. Annu. Rev. Pharmacol. Toxicol. **23**, 239–257 (1983)
5. Griendling, K.K., Sorescu, D., Ushio-Fukai, M.: NAD(P)H oxidase : role in cardiovascular biology and disease. Circul. Res. **86**, 494–501 (2000)
6. Pagano, P.J., et al.: An NADPH oxidase superoxide-generating system in the rabbit aorta. Am. J. Physiol. **268**, H2274–H2280 (1995)
7. Fridovich, I.: Superoxide dismutases. Annu. Rev. Biochem. **44**, 147–159 (1975)
8. Chance, B., Sies, H., Boveris, A.: Hydroperoxide metabolism in mammalian organs. Physiol. Rev. **59**, 527–605 (1979)
9. Fleming, I., Busse, R.: Molecular mechanisms involved in the regulation of the endothelial nitric oxide synthase. Am. J. Physiol. **284**, R1–R12 (2003)
10. Vásquez-Vivar, J., et al.: Superoxide generation by endothelial nitric oxide synthase: the influence of cofactors. Proc. Natl. Acad. Sci. USA **95**, 9220–9225 (1998)
11. Kelm, M., Dahmann, R., Wink, D., Feelisch, M.: The nitric oxide/superoxide assay. J. Biol. Chem. **272**, 9922–9932 (1997)
12. Pacher, P., Beckman, J.S., Liaudet, L.: Nitric oxide and peroxynitrite in health and disease. Physiol. Rev. **87**, 315–424 (2007)
13. Ducrocq, C., Blanchard, B.: Peroxynitrite: an endogenous oxidizing and nitrating agent. Cell. Mol. Life Sci. **55**, 1068–1077 (1999)
14. Ross, R.: The pathogenesis of atherosclerosis: a perspective for the 1990s. Nature **362**, 801–809 (1993)
15. Glass, C.K., Witztum, J.L.: Atherosclerosis: the road ahead. Cell **104**, 503–516 (2001)
16. Brown, M.S., Goldstein, J.L.: The SREBP pathway: regulation of cholesterol metabolism by proteolysis of a membrane-bound transcription factor. Cell **89**, 331–340 (1997)
17. Stocker, R., Keaney, J.F.: New insights on oxidative stress in the artery wall. J. Thromb. Haemost. **3**, 1825–1834 (2005)
18. Palinski, W., et al.: Low density lipoprotein undergoes oxidative modification in vivo. Proc. Natl. Acad. Sci. USA **86**, 1372–1376 (1989)
19. Nishi, K., et al.: Oxidized LDL in carotid plaques and plasma associates with plaque instability. Atert. Thromb. Vasc. Biol. **22**, 1649–1654 (2002)
20. Ehara, S., et al.: Elevated levels of oxidized low density lipoprotein show a positive relationship with the severity of acute coronary syndromes. Circulation **103**, 1955–1960 (2001)
21. Aviram, M., Fuhrman, B.: LDL oxidation by arterial wall macrophages depends on the oxidative status in the lipoprotein and in the cells: role of prooxidants vs. antioxidants. Mol. Cell. Biochem. **188**, 149–159 (1998)
22. Leeuwenburgh, C., et al.: Reactive nitrogen intermediates promote low density lipoprotein oxidation in human atherosclerotic intima. J. Biol. Chem. **272**, 1433–1436 (1997)
23. Kontush, A., Chapman, M.J.: Functionally defective high-density lipoprotein: a new therapeutic target at the crossroads of dyslipidemia, inflammation, and atherosclerosis. Pharmacol. Rev. **58**, 342–374 (2006)
24. Kontush, A., Chapman, M.J.: Antiatherogenic small, dense HDL—guardian angel of the arterial wall? Nat. Clin. Pract. Cardiovasc. Med. **3**, 144–153 (2006)
25. Aviram, M., Rosenblat, M.: Paraoxonases 1, 2, and 3, oxidative stress, and macrophage foam cell formation during atherosclerosis development. Free Radic. Biol. Med. **37**, 1304–1316 (2004)

# References

26. Nakajima, T., et al.: Characterization of the epitopes specific for the monoclonal antibody 9F5-3a and quantification of oxidized HDL in human plasma. Ann. Clin. Biochem. **41**, 309–315 (2004)
27. Zheng, L., et al.: Apolipoprotein A-I is a selective target for myeloperoxidase-catalyzed oxidation and functional impairment in subjects with cardiovascular disease. J. Clin. Invest. **114**, 529–541 (2004)
28. Francis, G.A.: High density lipoprotein oxidation: in vitro susceptibility and potential in vivo consequences. Biochim. Biophys. Acta Mol. Cell Biol. Lipids **1483**, 217–235 (2000)
29. Grundy, S.M., et al.: Implications of recent clinical trials for the national cholesterol education program adult treatment panel III guidelines. Circulation **110**, 227–239 (2004)
30. Lenfant, C.: Clinical research to clinical practice—lost in translation? N. Engl. J. Med. **349**, 868–874 (2003)
31. Steinberg, D., Glass, C.K., Witztum, J.L.: Evidence mandating earlier and more aggressive treatment of hypercholesterolemia. Circulation **118**, 672–677 (2008)
32. Waters, D.D., et al.: Predictors of new-onset diabetes in patients treated with atorvastatin: results from 3 large randomized clinical trials. J. Am. Coll. Cardiol. **57**, 1535–1545 (2011)
33. Preiss, D., et al.: Risk of incident diabetes with intensive-dose compared with moderate-dose statin therapy. JAMA J. Am. Med. Assoc. **305**, 2556–2564 (2011)
34. Rietjens, I.M.C.M., et al.: The pro-oxidant chemistry of the natural antioxidants vitamin C, vitamin E, carotenoids and flavonoids. Environ. Toxicol. Pharmacol. **11**, 321–333 (2002)
35. Fuhrman, B., Aviram, M.: Anti-atherogenicity of nutritional antioxidants. IDrugs **4**, 82–92 (2001)
36. Steinhubl, S.R.: Why have antioxidants failed in clinical trials? Am. J. Cardiol. **101**, 14D–19D (2008)
37. Bjelakovic, G., Nikolova, D., Gluud, L.L., Simonetti, R.G., Gluud, C.: Mortality in randomized trials of antioxidant supplements for primary and secondary prevention—Systematic review and meta-analysis. JAMA, J. Am. Med. Assoc. **297**, 842–857 (2007)
38. Gross, Z., Galili, N., Saltsman, I.: The first direct synthesis of corroles from pyrrole. Angew. Chem., Int. Ed. **38**, 1427–1429 (1999)
39. Mahammed, A., Goldberg, I., Gross, Z.: Highly selective chlorosulfonation of tris(pentafluorophenyl)corrole as a synthetic tool for the preparation of amphiphilic corroles and metal complexes of planar chirality. Org. Lett. **3**, 3443–3446 (2001)
40. Saltsman, I., et al.: Selective substitution of corroles: nitration, hydroformylation, and chlorosulfonation. J. Am. Chem. Soc. **124**, 7411–7420 (2002)
41. Haber, A., Aviram, M., Gross, Z.: Protecting the beneficial functionality of lipoproteins by 1-Fe, a corrole-based catalytic antioxidant. Chem. Sci. **2**, 295–302 (2011)
42. Kanamori, A., Catrinescu, M.M., Mahammed, A., Gross, Z., Levin, L.A.: Neuroprotection against superoxide anion radical by metallocorroles in cellular and murine models of optic neuropathy. J. Neurochem. **114**, 488–498 (2010)
43. Kupershmidt, L., et al.: Metallocorroles as cytoprotective agents against oxidative and nitrative stress in cellular models of neurodegeneration. J. Neurochem. **113**, 363–373 (2010)
44. Okun, Z., et al.: Manganese corroles prevent intracellular nitration and subsequent death of insulin-producing cells. ACS Chem. Biol. **4**, 910–914 (2009)
45. Haber, A., et al.: Amphiphilic/bipolar metallocorroles that catalyze the decomposition of reactive oxygen and nitrogen species, rescue lipoproteins from oxidative damage, and attenuate atherosclerosis in mice. Angew. Chem. Int. Ed. **47**, 7896–7900 (2008)
46. Agadjanian, H., et al.: Tumor detection and elimination by a targeted gallium corrole. Proc. Natl. Acad. Sci. USA **106**, 6105–6110 (2009)
47. Agadjanian, H., et al.: Specific delivery of corroles to cells via noncovalent conjugates with viral proteins. Pharm. Res. **23**, 367–377 (2006)
48. Aviv, I., Gross, Z.: Corrole-based applications. Chem. commun. (20), 1987–1999 (2007)
49. Gross, Z., Gray, H.B.: How do corroles stabilize high valent metals? Comments Inorg. Chem. **27**, 61–72 (2006)

50. Simonson, S.G., et al.: Aerosolized manganese SOD decreases hyperoxic pulmonary injury in primates. I. Physiology and biochemistry. J. Appl. Physiol. **83**, 550–558 (1997)
51. Salvemini, D., Wang, Z.-Q., Stern, M.K., Currie, M.G., Misko, T.P.: Peroxynitrite decomposition catalysts: therapeutics for peroxynitrite-mediated pathology. Proc. Natl. Acad. Sci. USA **95**, 2659–2663 (1998)
52. Batinić-Haberle, I., Rebouças, J.S., Spasojević, I.: Superoxide dismutase mimics: chemistry, pharmacology, and therapeutic potential. Antioxid. Redox Signal. **13**, 877–918 (2010)
53. Eckshtain, M., et al.: Superoxide dismutase activity of corrole metal complexes. Dalton Trans. (38), 7879–7882 (2009)
54. Mahammed, A., Gross, Z.: Highly efficient catalase activity of metallocorroles. Chem. Comm. **46**, 7040–7042 (2010)
55. Mahammed, A., Gross, Z.: Iron and manganese corroles are potent catalysts for the decomposition of peroxynitrite. Angew. Chem. Int. Ed. **45**, 6544–6547 (2006)
56. Lee, J., Hunt, J.A., Groves, J.T.: Manganese porphyrins as redox-coupled peroxynitrite reductases. J. Am. Chem. Soc. **120**, 6053–6061 (1998)

# Chapter 2
# Research Goals

1. To characterize the interactions between corroles and lipoproteins.
2. To explore the effect of corroles on oxidative and nitrosative damage to lipoproteins.
3. To characterize the interactions between corroles and macrophages.
4. To explore the effect of corroles on macrophage atherogenicity.
5. To investigate the effect of corroles on atherosclerosis development in mice.

# Chapter 3
# Results

## 3.1 Interactions of Corroles with Lipoproteins

### 3.1.1 Corrole-Lipoproteins Binding

LDL or HDL were incubated with 10 μM **1-Fe**, **1-Mn** or **1-Ga**, and the absorbance spectra were measured. The solutions were extensively dialyzed, and their absorbance spectra measured again. Final corrole concentrations in the solutions were calculated from the percent reduction in absorbance before and after dialysis, leading to the conclusion that the *maximal* binding capacity of the lipoproteins is 35–45 corrole molecules to each LDL particle and 10–12 to each HDL particle (Table 3.1).

### 3.1.2 Corrole Distribution in Serum

#### 3.1.2.1 Qualitative Analyses

Human serum was incubated with or without 40 μM **1-Fe**, **1-Mn** or **2-Fe** and then subjected to KBr density gradient separation (Fig. 3.1). This procedure caused fractionation of the serum, with the VLDL floating at the top of the tube, LDL forming a distinct yellow ring about a third tube length from the top, HDL providing a light yellow smear a third tube length from the bottom, and the lipoprotein-deficient serum (LPDS) residing at the bottom of the tube. For serum pre-treated with **1-Fe** or **1-Mn** (or **1-Ga**) the green color of the corrole was seen in the HDL fraction while the red **2-Fe** was located in the LPDS fraction.

**Table 3.1** The stoichiometry of corrole binding to isolated lipoproteins

| Lipoprotein type | Lipoprotein concentration (μM) | Initial corrole concentration (μM) | Absorbance reduction (%) | Final corrole concentration (μM) | Corrole molecules per lipoprotein |
|---|---|---|---|---|---|
| LDL | 0.2 | 10 | 10–30 | 7–9 | 35–45 |
| HDL | 0.6 | 10 | 25–35 | 6.5–7.5 | 10–12 |

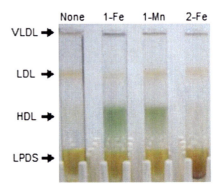

**Fig. 3.1** The serum distribution of **1-Fe**, **1-Mn** and **2-Fe** in a KBr density gradient. KBr fractionation of serum without or with 40 μM of **1-Fe**, **1-Mn** or **2-Fe**

### 3.1.2.2 Quantitative Analyses

Human serum containing 100 μM **1-Fe** (Fig. 3.2), **1-Mn** (Fig. 3.3) or **1-Ga** (Fig. 3.4) was subjected to size exclusion HPLC separation. Protein-containing fractions were detected by recording chromatograms at 280 nm, indicating elution of VLDL at 15 min, LDL at 23 min, HDL2 at 31 min, HDL3 at 34 min and non-lipoprotein associated proteins at more than 45 min. The presence of corrole in the various fractions was identified by 420 nm chromatograms together with full electronic spectra of the eluting fractions. Measurements of peak area from the 420 nm chromatograms revealed that 85 % of **1-Fe** was bound to HDL2, none was bound to HDL3, and the remaining 15 % was bound to LDL (the peak observed at the retention time of VLDL was due to other colored material, as it did not display the characteristic spectrum of **1-Fe**). **1-Mn** and **1-Ga** displayed lower binding selectivity, with the corrole residing in all lipoprotein fractions. About 60 % of these corroles was bound to HDL2, approximately 30 % to LDL and the remaining distributed both to HDL3 and to VLDL. The distribution of the aluminium and cobalt complexes of the bis-sulfonated corrole were also examined, but their binding was even less selective than that of **1-Mn** and **1-Ga**, and hence they were not further investigated. Similar HPLC experiments with the analogous sulfonated porphyrin **2-Fe** revealed that this compound did not elute with any of the serum constitutes under the very diluting conditions of the HPLC experiments, whereas the natural porphyrin hemin eluted almost exclusively within the HDL fraction (but in this case HDL3).

3.1 Interactions of Corroles with Lipoproteins

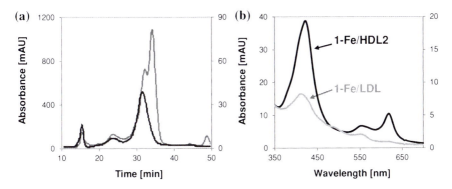

**Fig. 3.2** The serum distribution of **1-Fe** determined by HPLC. The chromatograms at 280 nm (*light line*, relative to the *left* y-axis) and 420 nm (*dark line*, relative to the *right* y-axis), obtained by HPLC separation of serum containing 100 μM **1-Fe** (**a**) and the absorbance spectra of the eluted **1-Fe**/HDL2 (relative to the *left* y-axis) and **1-Fe**/LDL (relative to the *right* y-axis) conjugates (**b**)

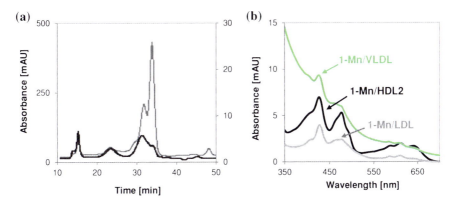

**Fig. 3.3** The serum distribution of **1-Mn** determined by HPLC. The chromatograms at 280 nm (*light line*, relative to the *left* y-axis) and 420 nm (*dark line*, relative to the *right* y-axis), obtained by HPLC separation of serum containing 100 μM **1-Mn** (**a**) and the absorbance spectra of the eluted **1-Mn**/HDL2, **1-Mn**/LDL and **1-Mn**/VLDL conjugates (**b**)

Examination of **1-Fe** distribution under high corrole concentrations was achieved by incubation of serum with 500 μM of **1-Fe** (while diluting the serum two fold, thus increasing the amount of corrole relative to serum ten folds relative to above mentioned experiments), its dialysis to remove any free or loosely bound corrole, and then its fractionation by the same HPLC method described above (Fig. 3.5). In this serum, containing its maximal load of **1-Fe**, the corrole was still bound only to lipoproteins, but with lower binding selectivity: 75 % to HDL2, 20 % to LDL, and the rest to VLDL and HDL3. The spectra of HPLC-eluted **1-Fe**/HDL2 conjugates for 100 μM (Fig. 3.2) relative to excess (Fig. 3.5) **1-Fe** were

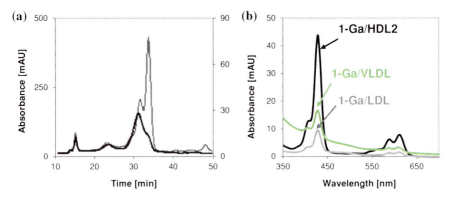

**Fig. 3.4** The serum distribution of **1-Ga** determined by HPLC. The chromatograms at 280 nm (*light line*, relative to the *left* y-axis) and 420 nm (*dark line*, relative to the *right* y-axis), obtained by HPLC separation of serum containing 20 μM **1-Ga** (**a**) and the absorbance spectra of the eluted **1-Ga**/HDL2, **1-Ga**/LDL and **1-Ga**/VLDL conjugates (**b**)

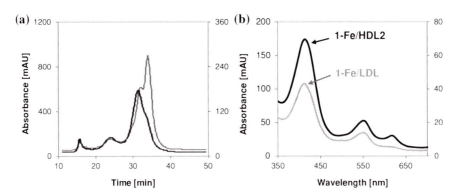

**Fig. 3.5** The serum distribution of excess **1-Fe** determined by HPLC. The chromatograms at 280 nm (*light line*, relative to the *left* y-axis) and 420 nm (*dark line*, relative to the *right* y-axis), obtained by HPLC separation of serum overloaded with **1-Fe** and dialyzed prior to separation (**a**) and the absorbance spectra of the eluted **1-Fe**/HDL2 (relative to the *left* y-axis) and **1-Fe**/LDL (relative to the *right* y-axis) conjugates (**b**)

somewhat different: (a) the Soret band differed by 6 nm in the two cases—424 nm relative to 418 nm for low and high corrole concentrations, respectively, and (b) the relative intensities of the two Q bands were dissimilar, with the 620 nm band [that does not appear in buffer solutions (Fig. 1.4, p. 11)] more intense for low corrole concentration and the 550 nm absorbance stronger for the high concentration. However, the characteristics of the **1-Fe**/LDL spectrum were not affected by the amount of corrole added to the serum. Examining the 280 nm/420 nm absorption ratio at 31 min (corresponding to the HDL2 peak maxima) and comparing it to the ratio measured for solutions in which known amounts of **1-Fe** were

3.1 Interactions of Corroles with Lipoproteins                                    21

added to purified HDL2 allowed for determination of the molar ratio between **1-Fe** and HDL2 within the conjugates. For the low **1-Fe** concentration, one corrole molecule per two HDL2 particles was calculated, whereas two corrole molecules per one HDL2 particle were identified when using the high **1-Fe** concentration.

The effect of different LDL and HDL concentrations in serum on the distribution of **1-Fe** and **1-Mn** was examined by adding 100 μM of corroles to sera with different lipid profiles, and conducting the above HPLC separation. Three very different sera were used, displaying normal lipoprotein values (LDL-C $\sim$ 100 mg/dL, HDL-C $\sim$ 50 mg/dL), high LDL and normal HDL values (LDL-C $\sim$ 200 mg/dL, HDL-C $\sim$ 50 mg/dL), and extremely low LDL but high HDL values (LDL-C $\sim$ 50 mg/dL, HDL-C $\sim$ 80 mg/dL). However, no significant differences in corrole distribution were displayed within the various serum samples.

### 3.1.2.3 Corrole Distribution in Serum of Treated Mice

Control C57Bl/6 mice were IP injected with 200 μL of 1 mM **1-Fe** or **1-Ga**, blood was taken from the mice 40 min later, and the serum was subjected to the HPLC separation (Figs. 3.6 and 3.7). The 420 nm chromatograph indicated the presence of colored species only within the HDL fraction, but the absorbance spectrum at the relevant time point was not precisely as expected for the corroles. In the case of **1-Fe** the characteristic 620 nm peak was observed, and for **1-Ga** the 610 nm peak was seen, indicating the presence of these compounds in the HDL fraction of the serum. In both cases additional long wavelength peaks appeared at 543 and 576 nm. Serum from non-treated mice also displayed the elution of a colored species at the same time point, that contained these two peaks but no peaks at a wavelength higher than 600 nm. A strong peak at 413 nm was also observed, indicating that this species was hemin, apparently liberated from red blood cells upon blood collection.

## 3.1.3 Spectral Changes in the Presence of Lipoprotein Components

### 3.1.3.1 Electronic Spectra with Lipoproteins

The spectra of the HPLC-eluted conjugates of **1-Fe** with HDL2 or LDL (Fig. 3.2, p. 19) and that in buffer (Fig. 1.4, p. 11) differed by the following aspects: (a) the Soret band of **1-Fe**/HDL2 was 10 nm red-shifted relative to that of **1-Fe**/LDL, which in turn was 10 nm red-shifted relative to the 404 nm band of **1-Fe** in buffer alone; and (b) only the former had a new distinct Q band at 620 nm, that also did not exist for **1-Fe** in buffer, which was stronger than the 550 nm band. These spectral differences were also reflected by the **1-Fe** buffer solutions turning from

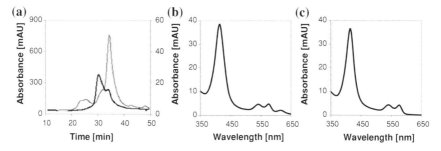

**Fig. 3.6** The in vivo serum distribution of **1-Fe** in mice. The chromatograms at 280 nm (*light line*, relative to the *left* y-axis) and 420 nm (*dark line*, relative to the *right* y-axis), obtained by HPLC separation of serum from C57Bl/6 mice IP injected with **1-Fe** (**a**) and the absorbance spectra of the fraction eluted at 31 min (**b**). The absorbance spectra of the fraction eluted at 31 min for the HPLC separation of serum from non-treated C57Bl/6 mice (**c**)

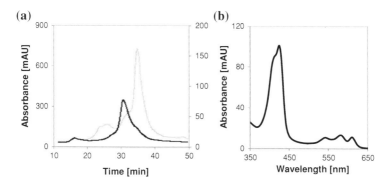

**Fig. 3.7** The in vivo serum distribution of **1-Ga** in mice. The chromatograms at 280 nm (*light line*, relative to the *left* y-axis) and 420 nm (*dark line*, relative to the *right* y-axis), obtained by HPLC separation of serum from C57Bl/6 mice IP injected with **1-Ga** (**a**) and the absorbance spectra of the fraction eluted at 31 min (**b**)

red to green when mixed with either whole serum or with isolated HDL2, but not when mixed with isolated LDL. **1-Mn** conjugates with both HDL and LDL displayed practically identical absorbance spectra (Fig. 3.3, p. 19), which differed from that in buffered solutions (Fig. 1.4, p. 11) by one main aspect: Upon addition of lipoproteins, the 480 nm band shifted to 476 nm and became the most intense absorbance peak. For **1-Ga**, no significant spectral changes were detected following lipoprotein binding (Fig. 3.4, p. 20; Fig. 1.4, p. 11).

### 3.1.3.2 CD Spectra with Lipoproteins

Circular dichroism (CD) spectra of corrole conjugates with HDL2 and LDL were measured. While **1-Fe**/LDL solutions provided no CD signals at the visible range, the

3.1 Interactions of Corroles with Lipoproteins

**Fig. 3.8** The CD spectra of 1-Fe conjugates with lipoproteins. The visible CD spectra of 1-Fe with HDL2 (*dark line*) or LDL (*light line*)

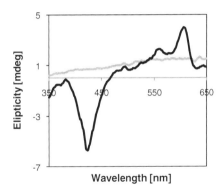

CD spectrum of **1-Fe**/HDL2 was very rich (Fig. 3.8). On the contrary, **1-Mn**/HDL2 and **1-Ga**/HDL2 solutions, as well as 1-Mn/LDL and **1-Ga**/LDL solutions, displayed no CD signals.

### 3.1.3.3 Electronic Spectra with Amino Acids

The electronic spectra of **1-Fe** in buffer following the addition of various basic amino acid esters were measured. The only amino acid that induced the formation of the **1-Fe**/HDL2 characteristic 620 nm band was histidine (Fig. 3.9). Titration of **1-Fe** with increasing amounts of the histidine ester displayed isosbestic points (at 410 and 590 nm) only at relatively low excess of histidine (up to ten fold excess), whereas the characteristic 620 nm band became more pronounced than that at 550 nm only under higher histidine concentrations ($\geq$40-fold excess). A spectrum very similar to that of the HPLC-eluted **1-Fe**/HDL2 conjugates (Fig. 3.2) was displayed only at very high histidine excess (>100 fold excess). Titration of **1-Mn** solutions with histidine displayed only very minor spectral changes and only at very high histidine excess (>100-fold).

### 3.1.3.4 Electronic Spectra with Liposomes

Liposomes (containing oleic acid phosphatidylcholine, cholesterol and cholesterol oleate) were reconstituted without proteins or with apolipoprotein E (apoE) or PON1, and their effect on the electronic spectra of **1-Fe** or **1-Mn** was examined. For **1-Fe** only minor changes were evident upon liposome addition (Fig. 3.10): the Soret band red-shifted by about 10 nm (relative to 404 nm in buffer alone) but without any change in the Q bands region (no 620 nm band). This change was similar to that obtained following LDL but not HDL2 addition to **1-Fe** (Fig. 3.2, p. 19). The spectrum of **1-Mn** was affected by either liposome solution at the same way it was affected by both LDL and HDL—a blue-shift and enhancement of the 480 nm band (Fig. 3.11; Fig. 3.3, p. 19).

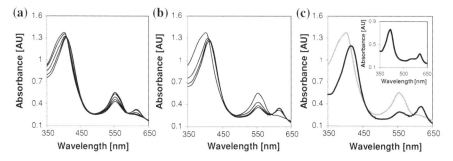

**Fig. 3.9** The effect of histidine on the absorbance spectrum of **1-Fe**. Changes in the absorbance spectrum of **1-Fe** upon its titration with histidine methyl ester: (**a**) 0, 2.5, 5.0 7.5, and 10.0 [histidine]/[**1-Fe**] ratio; (**b**) 0, 10, 20, and 30 [histidine]/[**1-Fe**] ratio; and (**c**) without or with 100-fold histidine excess (*grey* and *black lines*, respectively). Inset: The absorbance spectrum of **1-Fe** in the presence of a very large excess of histidine methyl ester

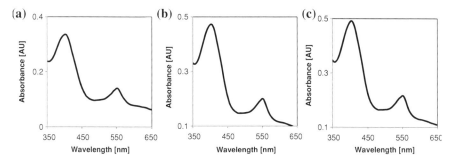

**Fig. 3.10** The absorbance spectra of **1-Fe** with liposomes. The absorbance spectra of **1-Fe** in the presence of liposomes containing only lipids (**a**) or also apoE (**b**) or PON1 (**c**)

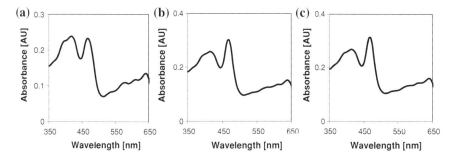

**Fig. 3.11** The absorbance spectra of **1-Mn** with liposomes. The absorbance spectra of **1-Mn** in the presence of liposomes containing only lipids (**a**) or also apoE (**b**) or PON1 (**c**)

3.1 Interactions of Corroles with Lipoproteins

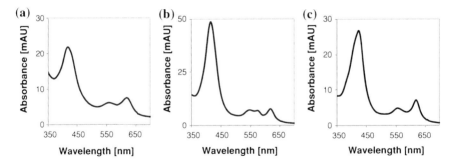

**Fig. 3.12** The spectra of **1-Fe**/HDL conjugates formed in the serum of various KO mice strains. The absorbance spectra of the fraction eluted at 31 min for the HPLC separation of serum from apoAI-KO (**a**), $E^0$ (**b**), and PON1-KO (**c**) mice treated with 100 μM **1-Fe**

#### 3.1.3.5 Electronic Spectra in Serum from PON1-KO and $E^0$ Mice

Serum was drawn from apoAI knock-out (apoAI-KO) mice apoE-deficient ($E^0$) mice, from PON1 knock-out (PON1-KO) mice, and from control mice (C57Bl/6), incubated with 100 μM **1-Fe**, and separated by HPLC. The corrole eluted within the HDL fraction for all mice types examined, and the eluted **1-Fe**/HDL displayed the 620 nm characteristic absorbance band in all cases (Fig. 3.12). In the case of the $E^0$-mice the spectrum was more complicated than expected for **1-Fe** as hemin co-eluted in this fraction (as also seen in the in vivo experiments, Fig. 3.7, p. 21).

#### 3.1.3.6 Electronic Spectrum with PON1-Treated HDL2

HDL2, presumed to be free of PON1, was incubated with PON1 at a concentration approximately that of apoAI, and the absorbance spectrum of **1-Fe** in such solutions was recorded. Three differences were displayed relative to non-PON1 treated solutions: the peak at 620 nm decreased very much, the 550 nm peak was enhanced, and the Soret band blue shifted by 5 nm (Fig. 3.13).

## 3.2 The Effect of Corroles on Oxidative-Stress-Induced Lipoprotein Damage

### 3.2.1 Copper-Ions-Induced Oxidation

LDL or HDL were incubated with or without increasing concentrations of **1-Fe**, **1-Mn** or **1-Ga**, and oxidation was initiated by the addition of copper sulfate (Fig. 3.14). For both lipoproteins, concentrations of ≥2.5 μM **1-Fe** completely eliminated

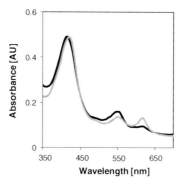

**Fig. 3.13** The effect of PON1 on the absorbance spectra of **1-Fe**/HDL2 conjugates. The absorbance spectra of **1-Fe** in HDL2 solutions that were pre-incubated with (*dark line*) or without (*light line*) PON1

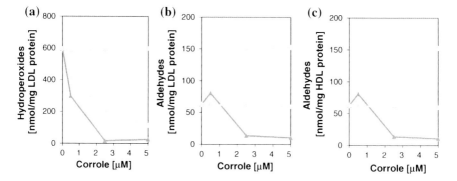

**Fig. 3.14** The dose-dependent effect of corroles on copper-ions-induced lipoprotein oxidation. The effect of various concentrations of **1-Fe** (▲), **1-Mn** (♦) or **1-Ga** (●) on LDL hydroperoxides (**a**); and aldehydes (**b**); formation 2 h after the addition of copper sulfate and on HDL aldehydes formation (**c**) 5 h after the addition of copper sulfate

hydroperoxides formation and reduced aldehydes formation by $\geq 80$ %. While **1-Mn** decreased hydroperoxides formation under all examined concentrations, a major dose-dependent increase in aldehydes formation was nevertheless observed. **1-Ga** had no effect on any of the examined parameters of lipoprotein oxidation.

Oxidation kinetics was measured for either LDL or HDL incubated with 2.5 μM of either **1-Fe** or **1-Mn**, followed by addition of copper sulfate (Fig. 3.15). In non-corrole treated LDL the appearance of oxidation products commenced after approximately 30 min from initiation, while HDL oxidation had virtually no delay time. When pre-treated with **1-Fe**, both LDL and HDL displayed no lipoprotein damage following addition of copper ions. However, in the presence of **1-Mn** oxidation was actually enhanced: (a) all examined oxidation products began to appear earlier than in control (with a delay time of only about 15 min for LDL); (b) conjugated dienes formation leveled off after only 50 min relative to 90 min in

## 3.2 The Effect of Corroles on Oxidative-Stress-Induced Lipoprotein Damage

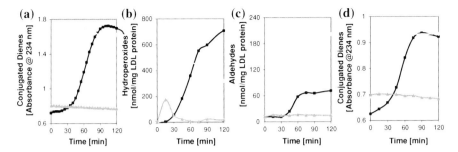

**Fig. 3.15** The effect of corroles on the kinetics of copper-ions-induced lipoprotein oxidation. Kinetics of conjugated dienes (**a**), hydroperoxides (**b**) and aldehydes (**c**) formation for LDL and conjugated dienes formation (**d**) for HDL following oxidation by copper sulfate without (■) or with 2.5 μM of **1-Fe** (▲) or **1-Mn** (♦)

the control; (c) the peak of hydroperoxides formation, which was not observed within the time scale of the experiment for the control, appeared after only 45 min; and (d) the amount of aldehydes formed was greatly increased relative to control.

### 3.2.2 Peroxynitrite-Induced Oxidation

#### 3.2.2.1 Establishing SIN-1 Working Concentration

HDL was incubated overnight with various concentrations of SIN-1, and then the amount of lipoprotein oxidation and nitration was examined. Lipid oxidation highly increased in correlation with the concentration of added SIN-1 (Fig. 3.16), as seen from the elevation in conjugated dienes and hydroperoxides levels. Only relatively low amounts of aldehydes were formed, reaching saturation at 250 μM of SIN-1. Protein oxidation was also highly dependent on SIN-1 concentrations (Fig. 3.17), with tryptophan fluorescence decreasing by 80 % with the highest examined concentration of SIN-1. Protein analysis by denaturating gel electrophoresis revealed several distinct differences between native and SIN-1-treated HDL associated proteins. For native HDL three main bands were observed: (a) near 72 kDa, corresponding to HSA (Mw = 66 kDa, identity confirmed by MS–MS); (b) near 26 kDa, corresponding to apoAI (Mw = 28 kDa, identity confirmed by MS–MS); and (c) at the front of the run (≤17 kDa), apparently corresponding to a variety of low molecular weight apolipoproteins. Following treatment with SIN-1 three dose-dependent changes were seen: (a) a new very high molecular weight band appeared, corresponding to apoAI oligomers (identified by MS–MS); (b) the migration distance of the monomeric apoAI band was shorter (reduced from $R_f = 0.69$ to $R_f = 0.66$); and (c) the intensity of the monomeric apoAI band was reduced. Another new band that was formed following SIN-1 treatment at about 43 kDa did not show a dependency on oxidant concentration. This band was found to contain

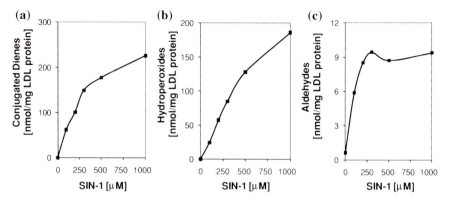

**Fig. 3.16** The dose-dependent effect of SIN-1 on lipoprotein lipid peroxidation. The effect of overnight treatment of HDL with various concentrations of SIN-1 on formation of fatty acid conjugated dienes (**a**), hydroperoxides (**b**) and aldehydes (**c**)

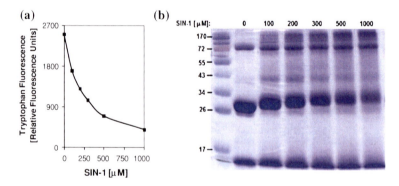

**Fig. 3.17** The dose-dependent effect of SIN-1 on lipoprotein protein oxidation. The effect of overnight treatment of HDL with various concentrations of SIN-1 on the oxidation of tryptophan residues (**a**) and on the electrophoretic pattern of HDL proteins (**b**)

apoAI (identified by MS–MS), and may thus be expected by its molecular weight to be a heterodimer of apoAI and some low molecular weight apolipoproteins, but supporting evidence for the presence of a second protein in the band were scarce.

A characteristic protein oxidative modification induced by RNS is nitration of tyrosine residues, a feature that was measured both by a competitive ELISA assay and by WB analysis (Fig. 3.18). While both methods revealed a large increase in total protein nitration in treatment with $\geq 500$ μM of SIN-1, the former was not sensitive enough to detect such modifications at lower concentrations (the absorbance of those samples was out of the range of the calibration curve, and so the negative values). Analysis of the specific bands on the WB membrane reveals that HDL-associated HSA was the first protein to be nitrated, a phenomenon which proceeds in a dose-dependent manner. Nitration of both monomeric and

3.2 The Effect of Corroles on Oxidative-Stress-Induced Lipoprotein Damage          29

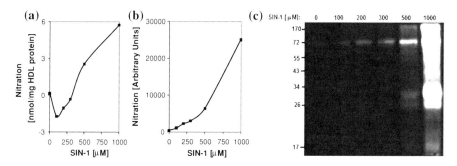

**Fig. 3.18** The dose-dependent effect of SIN-1 on lipoprotein protein nitration. The effect of overnight treatment of HDL with various concentrations of SIN-1 on nitration of tyrosine residues, measured by ELISA (**a**) and western blot analysis [densitometry (**b**) and membrane (**c**)]

oligomeric apoAI was only observed at SIN-1 concentration of 500 μM, and the nitro group was identified (by MS–MS) to be on the tyrosine at position 29 (while the new 43 kDa band was doubly nitrated at positions 29 and 18). At the highest SIN-1 concentration, specific nitrated bands were hard to identify, as a smear was seen thoroughout the whole lane.

### 3.2.2.2 The Effect of Corroles on Lipoprotein Oxidation

HDL was incubated with or without increasing concentrations of either **1-Fe** or **1-Mn**, and then SIN-1 was added and lipoprotein oxidation (Fig. 3.19) and nitration (Fig. 3.20) were measured. Inspection of conjugated dienes formation indicated ≥25 μM **1-Fe** to partially prevent lipid oxidation, while **1-Mn** had apparently no effect. Kinetic measurements at 50 μM corrole validated the results for **1-Fe**, with a 70 % decrease in the level of conjugated dienes, but elucidated that **1-Mn** actually affected oxidation in a negative sense—it reduced the oxidation delay time (without a major change in the total amount of conjugated dienes, as also seen in the copper ions case). Quenching of tryptophan fluorescence could not be measured in the presence of the corrole, as their presence significantly reduced the emission. Examination of the electrophoretic pattern of HDL-associated proteins showed that at low **1-Fe** concentrations the intensity of the oligomeric apoAI band has increased and the migration distance of the monomeric apoAI was reduced, corresponding to higher oxidation levels. Both phenomena gradually decreased by increasing the concentration of the corrole to 50 μM (and completely disappeared at a concentration of 100 μM). On the other hand, **1-Mn** dose-dependently elevated the intensity of the apoAI oligomers band and decreased the intensity and the migration distance of the apoAI monomers band. **1-Fe** also reduced protein nitration levels already at the lowest applied concentration, and completely prevented it from a concentration of ≥25 μM (giving a nitration level identical to that seen for native HDL). On the other hand, **1-Mn** caused a pronounced increase in

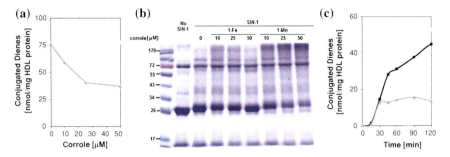

**Fig. 3.19** The effect of corroles on SIN-1-induced HDL oxidation. The effect of various concentrations of **1-Fe** (▲) or **1-Mn** (♦) on HDL conjugated dienes formation (a) and on the electrophoretic pattern of HDL proteins (b) after 2 h from the addition of SIN-1, and the effect of 50 µM of the corroles on the kinetics of HDL conjugated dienes formation (c)

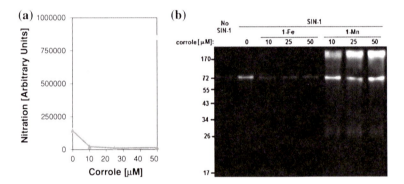

**Fig. 3.20** The effect of corroles on SIN-1-induced HDL nitration. The effect of various concentrations of **1-Fe** (▲) or **1-Mn** (♦) on nitration of HDL tyrosine residues, measured by WB analysis [densitometry (a) and membrane (b)]

the amount of nitration of both monomeric and oligomeric apoAI under all investigated concentrations.

## 3.2.3 Lipoprotein Function

### 3.2.3.1 Anti-Atherogenic Functions of HDL

HDL was treated with or without 50 µM of **1-Fe**, and then exposed to vehicle, copper ions or SIN-1. After overnight incubation the variously treated samples were examined for three important anti-atherogenic functions of HDL (Fig. 3.21): cholesterol efflux capacity, antioxidant activity and anti-apoptotic capability. Treatment by either SIN-1 or copper ions reduced HDL-mediated cholesterol

3.2 The Effect of Corroles on Oxidative-Stress-Induced Lipoprotein Damage       31

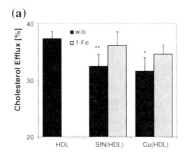

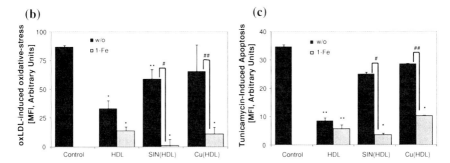

**Fig. 3.21** The effect of **1-Fe** on oxidative-stress induced loss of anti-atherogenic HDL functions. The effect of 50 μM of **1-Fe** on HDL loss of function that follows overnight oxidation with SIN-1 or copper ions: cholesterol efflux capacity (**a**), inhibition of oxLDL-induced oxidative-stress (**b**), and inhibition of tunicamycine-induced apoptosis (**c**) in macrophage cells. $*p < 0.005$, $**p < 0.05$ (vs. control). $^{\#}p < 0.005$, $^{\#\#}p < 0.05$ (vs. w/o corrole)

efflux from macrophage cells by 14 and 20 % relative to native HDL, but only by 3 and 8 % for oxHDL pre-treated with **1-Fe**, respectively. Oxidative-stress induction in macrophages by oxLDL treatment was inhibited by: 60 % in the presence of native HDL, 25–30 % with SIN-1 or copper ions oxidized HDL, and 85 % or more for *each* sample containing **1-Fe**/HDL conjugates (with or without subsequent oxidation). Tunicamycin-induced apoptosis of macrophages was inhibited by 75 % with native HDL, by only 15–30 % for oxHDL, but 70–90 % by all **1-Fe**/HDL conjugates.

### 3.2.3.2 Pro-Oxidative Activity of oxLDL

LDL was heavily oxidized by exposure to copper ions, and only then incubated overnight with 50 μM **1-Fe**. The samples were then added to macrophages, and the resulting increase in the oxidative status of the cells was measured. Cell oxidation for oxLDL treated with **1-Fe** was 55 % lower than that in cells exposed to non-corrole-treated oxLDL (Fig. 3.22).

**Fig. 3.22** The effect of **1-Fe** on the pro-oxidative activity of oxLDL. Oxidative status of macrophages treated by oxLDL that was pre-treated overnight with or without 50 μM of **1-Fe**. $^{\#\#}p < 0.05$ (vs. w/o corrole)

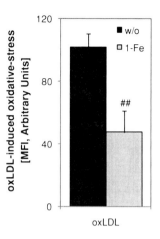

## 3.3 Interactions of Corroles with Macrophages

### 3.3.1 Fluorescence-Based Detection Of 1-Ga

#### 3.3.1.1 Detection by Various Methods

J774.A1 macrophage cells were incubated with or without 20 μM of **1-Ga** for 30 min, washed, fixed, stained with DAPI, and examined by fluorescence microscopy. In treated cells the red emission of the corrole was clearly seen, and together with the blue nuclear staining **1-Ga** was identified to reside in the cytoplasm and not in the nucleus (Fig. 3.23). These experiments also confirmed that the corrole was not decomposed following its uptake into the cells, as **1-Ga** fragments are not fluorescent.

J774.A1 macrophage cells were incubated with 0, 5 or 20 μM of **1-Ga** for 30 min, washed and analyzed by flow cytometry (Fig. 3.24). While control cells displayed a unified population of cells, **1-Ga**-treated cells showed two distinct populations, with the mean fluorescence intensity of each population increasing in correlation with increased **1-Ga** concentration. However, the high intensity population of cells was characterized by small size and low granularity, indicative that these were actually dead cells.

The fluorescence of various concentrations of **1-Ga** under different conditions was measured for creating calibration curves. The fluorescence intensity was found to be very much dependant on the contents of the media in which it was measured, with the presence of proteins increasing the fluorescence very much (Fig. 3.25). To allow for measurement of cellular-derived corrole, calibration curves were measured in the presence of cellular debris (which was achieved by culturing J774.A1 macrophage cells in the examination plate, suspending them in distilled water and then treating them with a freeze–thaw cycle). Basal fluorescence was

3.3 Interactions of Corroles with Macrophages 33

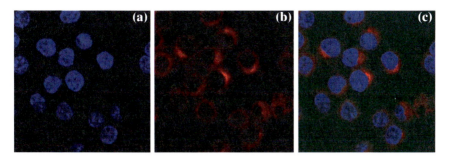

**Fig. 3.23** Microscopic detection of **1-Ga** in macrophage cells. The cellular localization of **1-Ga** in macrophages: nuclear staining (**a**), corrole fluorescence (**b**) and the merge of the two (**c**)

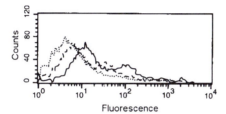

**Fig. 3.24** Flow cytometry detection of **1-Ga** in macrophage cells. Cellular fluorescence distribution following treatment of the cells with 0 μM (*dotted line*), 5 μM (*dashed line*) and 20 μM (*full line*) of **1-Ga**

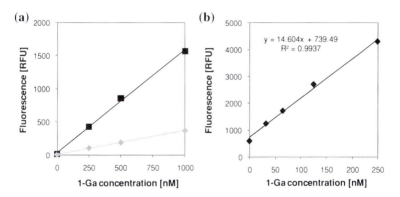

**Fig. 3.25** Calibration curves for quantification of cellular-derived **1-Ga**. Calibration curves of **1-Ga** in: (**a**) cell culture media with (*dark line*) or without (*light line*) FCS; and (**b**) in water containing cellular debris

displayed by the cells, and addition of **1-Ga** increased fluorescence in a dose-dependent and linear fashion. This calibration curve allowed for the quantification of intracellular **1-Ga**.

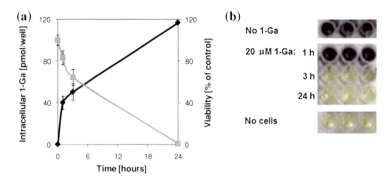

**Fig. 3.26** Time-dependent cellular accumulation of **1-Ga**. (**a**) The amount of **1-Ga** accumulating in macrophages (*dark line*, relative to the *left* y-axis) and its effect on cellular viability (*light line*, relative to the *right* y-axis). (**b**) The wells of the viability test—the yellow dye used in the assay turns purple in the presence of living cells

### 3.3.1.2 Time-Dependent Intracellular Accumulation of 1-Ga

J774.A1 macrophages were incubated with 20 μM of **1-Ga** for various times, washed to remove non-internalized corrole, and analyzed for viability or for intracellular concentrations of **1-Ga** (Fig. 3.26). The latter was achieved by rupture of the cells, fluorescence measurement, and concentration determination relative to the standard curve (reaction volume was 100 μL, so 100 nM indicated from the calibration curve was translated to 10 pmol/well). Cellular accumulation of **1-Ga** was very rapid in the first hour of incubation, and continued even up to 24 h of incubation, but in a more moderate fashion. However, this accumulation was accompanied by reduction in cell viability, with only 65 % of the cells surviving 3 h of incubation, and complete cell death after 24 h. At **1-Ga** concentration of 2 μM the compound displayed much lower intracellular levels (an order of magnitude less) and cellular viability was not decreased. This was in line with $IC_{50}$ values, found to be in the range of 10–50 μM, depending on the incubation time (4 to 48 h).

## 3.3.2 Chemiluminescence-Based Detection of 1-Fe

### 3.3.2.1 A New Detection Method

**1-Fe**, luminol and $H_2O_2$ were mixed, and emission at 430 nm was continuously measured for 6 min. Intense chemiluminescence was displayed down to 10 nM of **1-Fe** at pH = 13 (Fig. 3.27) and pH = 9 (Fig. 3.28), whereas 1,000 nM corrole were needed for obtaining curves with a good signal to noise ratio at pH = 7 (Fig. 3.29). The characteristics of all examined reactions depend on the amount of added $H_2O_2$ relative to that of luminol (1 mM): for equimolar ratios the initial

3.3 Interactions of Corroles with Macrophages

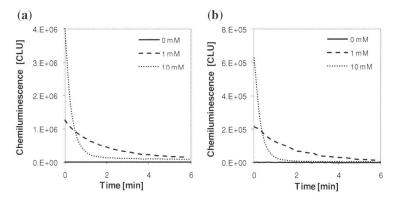

**Fig. 3.27** **1-Fe** catalyzed luminol emission in cell-free systems at pH = 13. Emission kinetics of solutions containing luminol, various $H_2O_2$ concentrations (see legend) and 100 nM (**a**) or 10 nM (**b**) **1-Fe** at pH = 13

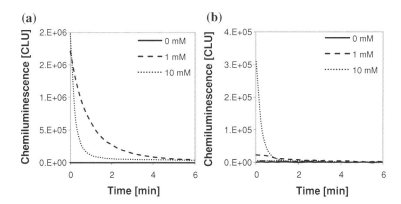

**Fig. 3.28** **1-Fe** catalyzed luminol emission in cell-free systems at pH = 9. Emission kinetics of solutions containing luminol, various $H_2O_2$ concentrations (see legend) and 100 nM (**a**) or 10 nM (**b**) **1-Fe** at pH = 9

**Fig. 3.29** **1-Fe** catalyzed luminol emission in cell-free systems at pH = 7. Emission kinetics of solutions containing luminol, various $H_2O_2$ concentrations (see legend) and 1,000 nM **1-Fe** at pH = 7

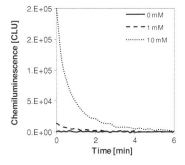

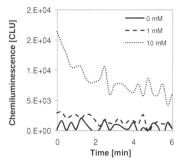

**Fig. 3.30** **1-Mn** catalyzed luminol emission in cell-free systems at pH = 13. Emission kinetics of solutions containing luminol, various $H_2O_2$ concentrations (see legend) and 100 nM **1-Mn** at pH = 13

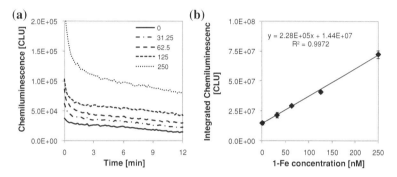

**Fig. 3.31** Calibration curve for quantification of cellular-derived **1-Fe**. (**a**) The emission kinetics upon addition of luminol, $H_2O_2$ and various concentrations of **1-Fe** to cellular debris. (**b**) The calibration curve obtained from the integration of the kinetic curves

luminescence was lower and the kinetics was slower than when using a ten-fold excess of $H_2O_2$. No emission was recorded for addition of **1-Fe** and luminol in the absence of $H_2O_2$. The same procedure was applied for **1-Mn**, however even in basic solutions (pH = 13) and corrole concentration of 100 nM, only very low chemiluminescence was induced by this compound (Fig. 3.30).

**1-Fe**, luminol and $H_2O_2$ were added to ruptured J774.A1 macrophages, and emission at 430 nm was continuously measured for 12 min (Fig. 3.31). In non-treated cell debris, addition of luminol and $H_2O_2$ (1 mM and 10 mM respectively, pH = 13) produced weak chemiluminescence. Upon addition of nanomolar concentrations of **1-Fe** to the ruptured cells, the emission was strongly enhanced in a dose-dependent manner. Integration of the kinetic curves allowed for the construction of a calibration curve, enabling the detection and quantification of cellular-derived **1-Fe** down to a concentration of 30 nM.

J774.A1 macrophage cells were incubated with or without 20 μM of **1-Fe**, washed and then ruptured or fixed. Next, luminol and $H_2O_2$ were added under

3.3 Interactions of Corroles with Macrophages

basic conditions (1 and 10 mM respectively, pH = 13), and chemiluminescence was measured. The two main observations that were: (a) chemiluminescence was displayed by both ruptured and fixed cells (with high standard deviation values in the latter case), indicating that fixation does not interfere with the chemiluminescent reaction; and (b) when fixed the cells remained intact (as visualized by microscopic examination) even though they were exposed to alkaline conditions. These results indicated that the new method may be applied for identifying the location of **1-Fe** in whole cells by microscopy. Accordingly, J774.A1 macrophage cells were incubated with or without 20 μM of **1-Fe**, washed and fixed, and then emission following the addition of luminol and $H_2O_2$ was examined by fluorescent microscopy (without excitation). However, no emission was detected, probably because more sensitive equipment and more sophisticated software were needed.

### 3.3.2.2 Time- and Dose-Dependent Intracellular Accumulation Of 1-Fe

J774.A1 macrophages were incubated with 20 μM of **1-Fe** for various times, washed to remove non-internalized corrole, and analyzed for viability or for intracellular amounts of **1-Fe** (Fig. 3.32). The latter was achieved by rupture of the cells, chemiluminescence measurement, and concentration determination relative to the standard curve (reaction volume was 100 μL, so 100 nM indicated from the calibration curve is translated to 10 pmol/well). Cellular accumulation of **1-Fe** was rapid in the first 4 h of incubation, and increased even after 24 h of incubation, but in a more moderate fashion. This accumulation was actually accompanied by an increase in cell viability of up to 120 % after 24 h. **1-Fe** was found to display a negative effect on cell survival only for prolonged incubation of high concentrations (a reduction to 80 % viability after 48 h with 100 μM).

J774.A1 macrophages were incubated with 20 μM of **1-Fe** for 24 h, washed to remove non-internalized corrole, further incubated without corrole in the culture media, and analyzed for intracellular amounts of **1-Fe** (Fig. 3.32). Intracellular levels of the corrole were reduced in correlation with the time of incubation in its absence, showing clearance of the corrole from the cells; however **1-Fe** was still detected in the cells after 24 h. Cell survival was high and did not significantly changed throughout the incubation time.

Increasing concentrations of **1-Fe** were incubated with J774.A1 macrophage cells for 2 h, followed by cell rupture and chemiluminescence detection (Fig. 3.33). The chemiluminescence intensity increased as a function of the added concentration of **1-Fe**, thus signaling an increase in the level of intracellular corrole, and reached saturation at 100 μM extracellular corrole. Quantification of the luminescence kinetics relative to a standard curve showed that intracellular **1-Fe** concentration started to level off at about 100 pmol/well.

Attempts were made to identify factors affecting accumulation of **1-Fe** in macrophages. Comparing cellular uptake with or without serum in the medium (all the above experiments were conducted in the presence of serum) revealed higher corrole uptake in the absence of serum; however, this was accompanied by

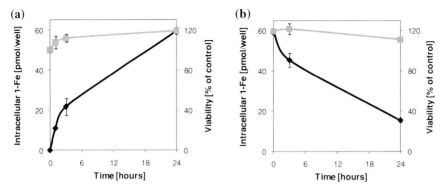

**Fig. 3.32** Time-dependent cellular accumulation of **1-Fe**. (**a**) The amount of **1-Fe** accumulating in macrophages (*dark line*, relative to the *left* y-axis) and its effect on cellular viability (*light line*, relative to the *right* y-axis). (**b**) The amount of **1-Fe** remaining in macrophages following 24 h corrole loading and further incubation without corrole in the cell media (*dark line*, relative to the *left* y-axis) and its effect on cellular viability (*light line*, relative to the *right* y-axis)

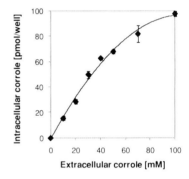

**Fig. 3.33** Dose-dependent cellular accumulation of **1-Fe**. The amount of **1-Fe** accumulating in macrophages following 2 h incubation with various corrole concentrations

reduction of cell survival. The higher cell accumulation can thus be attributed to increased membrane permeability that accompanies cell death. The effect of individual serum components was examined by supplementing the culture medium with only HDL, only LDL or only BSA (without serum). In all cases **1-Fe** was taken up by the cells, but the experiments did not allow for determining which serum component was the most effective in mediating corrole uptake to the cells.

### 3.3.3 Cellular Accumulation of 1-Fe Versus Iron(III) Porphyrins

**2-Fe**, **3-Fe** or hemin together with luminol and $H_2O_2$ were added to ruptured J774.A1 macrophages, and emission at 430 nm was continuously measured for 12 min (Fig. 3.34). Integration of the kinetic curves allowed for the construction of calibration curves, as was previously described for **1-Fe**. The chemiluminescence

## 3.3 Interactions of Corroles with Macrophages

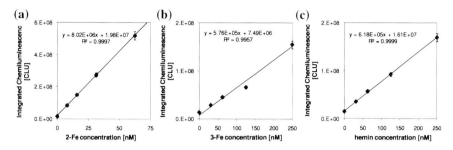

**Fig. 3.34** Calibration curve for quantification of cellular-derived **2-Fe**, **3-Fe**, and hemin. The calibration curve obtained from the integration of the kinetic curves for the chemiluminescence reaction in the presence of **2-Fe** (**a**), **3-Fe** (**b**), and hemin (**c**)

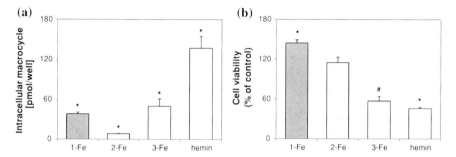

**Fig. 3.35** Cellular accumulation of **1-Fe**, **2-Fe**, **3-Fe**, and hemin and consequential cell survival. The amount of **1-Fe**, **2-Fe**, **3-Fe**, and hemin accumulating in macrophages after 24 h incubation (**a**) and their effect on cellular viability (**b**). *$p < 0.001$ and #$p < 0.05$ relative to non-treated cells

was more intense when the reactions were catalyzed by these complexes rather than by **1-Fe**, with **2-Fe** being extremely efficient, as even 7 nM were easily detected (and the curve slope was more than ten fold higher for **2-Fe**).

J774.A1 macrophage cells were incubated with 20 μM of **1-Fe**, **2-Fe**, **3-Fe** or hemin in serum-containing medium for 24 h, followed by removal of medium, cell rupture and determination of intracellular concentrations relative to standard curves; and cellular viability was analyzed in parallel (Fig. 3.35). **2-Fe** accumulated to only about 8 pmol/well, 40–50 pmol/well were obtained for **1-Fe** and **3-Fe**, and 140 pmol/well for hemin. However, major differences in cellular viabilities were seen: **1-Fe** elevated cell viability to 145 %, **2-Fe** had no significant effect, and **3-Fe** and hemin reduced cell survival to 45–60 %.

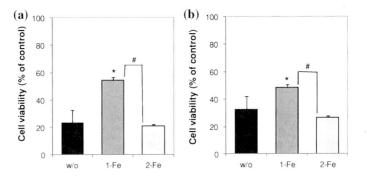

**Fig. 3.36** The effect of **1-Fe** and **2-Fe** against cellular death induced by primary oxidants. The effect of **1-Fe** and **2-Fe** on $H_2O_2$ (**a**) and SIN-1 (**b**) induced cellular death

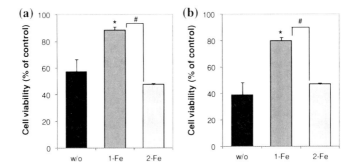

**Fig. 3.37** The effect of **1-Fe** and **2-Fe** against cellular death induced by indirect oxidants. The effect of **1-Fe** and **2-Fe** on oxLDL (**a**) and LPS + INF-$\gamma$ (**b**) induced cellular death

## 3.4 The Effect of Corroles on Macrophage Atherogenicity

### 3.4.1 Effect on Oxidative-Stress-Induced Cell Death

J774.A1 macrophages were incubated without (w/o) or with 20 μM of **1-Fe** or **2-Fe** for 24 h. Then cells were washed and treated without (control) or with: $H_2O_2$, SIN-1, oxLDL or LPS + INF-$\gamma$, followed by determination of cellular viability (Figs. 3.36 and 3.37). **2-Fe** had no significant effect in any of the examined cases, but the 40 pmol/$10^5$ cells intracellular **1-Fe** increased cell survival for all applied toxins: from 25 to 55 % for $H_2O_2$, from 35 to 50 % for SIN-1, from 55 to 90 % for oxLDL, and from 40 to 80 % for LPS + INF-$\gamma$. Interestingly, **1-Mn** was almost as efficient as **1-Fe** in protecting the cells against all examined toxins.

To gain additional insight, the effect on cellular oxidation and nitration were analyzed as well. While cellular oxidation was largely increased following the

3.4 The Effect of Corroles on Macrophage Atherogenicity   41

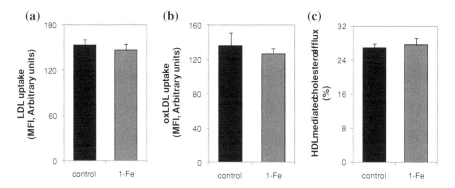

**Fig. 3.38** The effect of **1-Fe** on macrophage cholesterol flux. The effect of **1-Fe** on LDL uptake (**a**), oxLDL uptake (**b**), and HDL-mediated cholesterol efflux (**c**)

various treatments (as determined by flow cytometry following addition of DCFH), the effect of the corroles could not be examined, as they act as catalysts of the detection reaction (See Appendix). Regarding cellular nitration, the basal levels in the macrophages (as determined by WB analysis) was so high, that no significant effect was displayed by the various toxins.

### 3.4.2 Effect on Macrophages Cholesterol Metabolism

J774.A1 macrophages were incubated with 20 µM of **1-Fe** for 24 h, washed to remove non-internalized corrole, and analyzed for LDL and oxLDL uptake by the cells, as well as for HDL-mediated cholesterol efflux from the cells (Fig. 3.38). Cellular loading with **1-Fe** did not affect any of these cholesterol trafficking pathways.

J774.A1 macrophages were incubated with increasing concentrations of **1-Fe** for 24 h, washed to remove non-internalized corrole, and analyzed for cellular cholesterol biosynthesis and for intracellular amounts of **1-Fe** (Fig. 3.39). Cholesterol biosynthesis was assayed by measuring incorporation of radiolabelled acetate into cholesterol following overnight starvation of the cells as to up-regulate the levels of HMGCR and increase cholesterol formation; and **1-Fe** intracellular concentration was determined by the new chemiluminescent method. While only about 1 % of the originally added **1-Fe** remained within the cells at the time of the biochemical assay (when adding 20 µM of **1-Fe**, 20 pmol corrole per $10^5$ cells were measured in 100 µL, which equals 200 nM of corrole), this was enough to significantly reduced cholesterol biosynthesis by the cells. The amounts of newly synthesized cholesterol by the macrophages decreased as the dosage of **1-Fe** increased, in parallel with an increase in **1-Fe** cellular uptake, indicating a dose dependent effect of the corrole on inhibition of cholesterol biosynthesis.

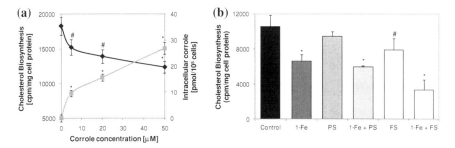

**Fig. 3.39** The effect of **1-Fe** cholesterol biosynthesis rate in macrophages. (**a**) The amount of **1-Fe** accumulating in macrophages (*light line*, relative to the *right* y-axis) and its effect on cellular cholesterol biosynthesis (*dark line*, relative to the *left* y-axis). (**b**) The effect of **1-Fe**, pravastatin (PS) or fluvastatin (FS), alone or in combination, on cellular cholesterol biosynthesis. *$p < 0.001$ relative to control

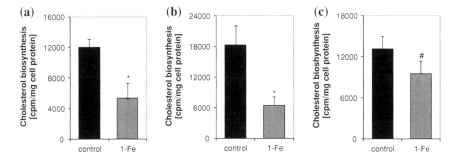

**Fig. 3.40** The effect of **1-Fe** administration to mice on macrophage cholesterol biosynthesis rate. The effect of **1-Fe** on cholesterol biosynthesis in macrophages harvested from young male (**a**) and female (**b**) and aged male (**c**) mice treated with the corrole. *$p < 0.01$ relative to control. #$p < 0.05$ relative to control

J774.A1 macrophages were incubated for 24 h with or without: 20 μM **1-Fe**, 200 μM pravastatin (PS), 20 μM fluvastatin (FS), or a combination of corrole and statin. Cholesterol biosynthesis was determined after removal of the non-internalized fraction of the tested compounds and overnight starvation (Fig. 3.39). The reduction in biosynthesis of cholesterol was determined to be 35 % for **1-Fe** alone, 10 % for pravastatin alone, 45 % for **1-Fe** with pravastatin, 25 % for fluvastatin alone and 70 % for **1-Fe** in combination with fluvastatin.

Young (12 weeks of age, divided to male and female groups) and aged (male, 30 weeks of aged) $E^0$-mice were orally treated with 10 mg/kg/day **1-Fe** for 12 weeks, and then macrophages were harvested from their peritoneum and examined for cholesterol biosynthesis ability (Fig. 3.40). **1-Fe** inhibited cholesterol biosynthesis by 50 % in male and 60 % in female young mice and by 30 % in aged mice relative to respective controls. In all cases the reduction in cholesterol biosynthesis following **1-Fe** administration was accompanied by a decrease in the

3.4 The Effect of Corroles on Macrophage Atherogenicity

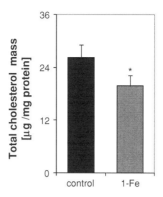

**Fig. 3.41** The effect of **1-Fe** administration to mice on macrophage cholesterol content. The effect of **1-Fe** on total cholesterol mass in macrophages harvested from aged male mice treated with the corrole. *$p < 0.01$ relative to control

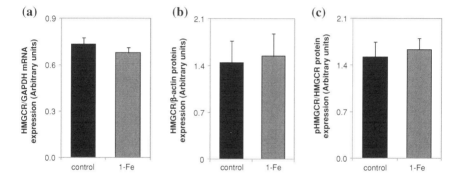

**Fig. 3.42** The effect of **1-Fe** on HMGCR expression. The effect of **1-Fe** on HMGCR mRNA expression (**a**), protein expression (**b**) and protein phosphorylation (**c**) levels

total amount of cholesterol within the cells, with the most significant decline of 25 % exhibited by the aged mice (Fig. 3.41).

The effect of **1-Fe** on cholesterol biosynthesis from mevalonate, a substrate downstream to the rate determining enzyme HMGCR, was examined. **1-Fe** displayed no effect, indicating that it performed its action on an earlier stage. This may point toward the HMGCR catalyzed reaction as the point at which **1-Fe** may act. However, **1-Fe** did not affect RNA or protein expression of HMGCR, nor did it affect the phosphorylation level of the enzyme, a factor that controls enzyme activity (Fig. 3.42). Examination of the HMGCR reaction in a cell-free system showed, that while pravastatin effectively inhibited the reaction, **1-Fe** seemed to catalyse it, as the consumption of NADPH was enhanced (Fig. 3.43).

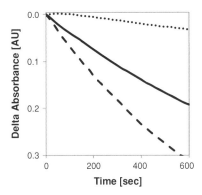

**Fig. 3.43** The effect of **1-Fe** on HMGCR activity in a cell-free system. The kinetics of NADPH disappearance from solution containing HMG-CoA, NADPH and HMGCR alone (*full line*), or in addition of **1-Fe** (*dashed line*) or pravastatin (*dotted line*)

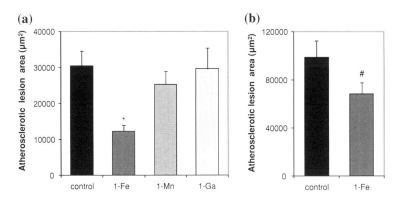

**Fig. 3.44** The effect of corroles on atherosclerosis development in young and aged mice. The effect of **1-Fe**, **1-Mn** and **1-Ga** on atherosclerotic lesion area in young mice (**a**) and of **1-Fe** in aged mice (**b**). *$p < 0.01$ relative to control. #$p < 0.05$ relative to control

## 3.5 The Effect of Corroles on the Development of Atherosclerosis

Young (12 weeks of age) and aged (30 weeks of aged) $E^0$-mice were orally treated with 10 mg/kg/day corrole, sacrificed and then the area of the lesion in their aortic arch was measured (Fig. 3.44). **1-Fe** significantly reduced plaque area by 60 and 30 % for young and aged mice, respectively. **1-Mn** displayed only a 16 % decrease in lesion area in the young mice, while **1-Ga** had no effect.

# Chapter 4
# Discussion

## 4.1 Interaction of Corroles with Lipoproteins

### 4.1.1 Corrole Distribution in Serum

Biodistribution of medications is a major factor dictating therapy efficacy, as demonstrated by many very excellent in vitro performing drugs displaying limited in vivo efficiency [1–5]. Every therapeutic first encounters the serum components, which may bind it and serve as carriers for its delivery within the body. Accordingly, serum distribution of potential drugs is a main issue of their activity, and the first step disclosed in this research. Earlier studies have focused on the interaction of the bis-sulfonated corrole metal complexes with *isolated* serum proteins. These investigations revealed very strong binding of the corroles to the most abundant serum proteins transferrin and albumin (with dissociation constants of $10^{-7}$–$10^{-9}$ and $>10^{-9}$ M, respectively) [6, 7]. Examinations of corrole interaction with *isolated* lipoproteins revealed high affinity binding in this case as well, and a binding stoichiometry of about 40 corrole molecules to *isolated* LDL and 10 to *isolated* HDL[1] (Table 3.1, p. 18). Together these results underline the need to directly explore the distribution of the corroles in whole serum, as strong binding may be outcompeted by stronger binding, and because the equilibrium between the various serum components depends also on their relative concentrations.

Examination of the distribution of **1-Fe**, **1-Mn** and **1-Ga** in serum, both by density gradient and by HPLC, revealed that these compound bind only to lipoproteins and not to "free" proteins. The highest percentage of corrole was found in HDL2 (Fig. 4.1): 85 % for **1-Fe** and 60 % for the other two corroles. 30 % of **1-Mn** and **1-Ga** were bound to LDL and the remaining 10 % to VLDL and HDL3, but only 15 % of **1-Fe** was bound to LDL and none of it to the two other lipoprotein subclasses. Lipoproteins are constructed of a hydrophobic core, in which triglycerides and cholesterol esters are found, surrounded by an amphipolar coat,

---

[1] HDL contains albumin (even when isolated), as may be seen from the SDS-PAGE experiments.

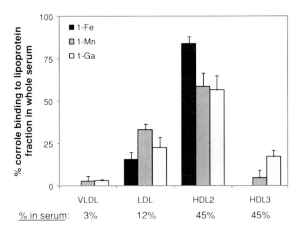

Fig. 4.1 Serum distribution of **1-Fe**, **1-Mn** and **1-Ga**. Percentage of corroles in each lipoprotein fraction, as determined from the HPLC experiments, and percentage of the lipoprotein fractions in serum

made of phospholipids and unesterified cholesterol. The bis-sulfonated corrole is also amphipolar, with negative charges on only one side of the otherwise hydrophobic molecule, which explains its high binding affinity to lipoproteins (Fig. 3.1, p. 18). This spontaneous association—strong enough to withstand the very diluting chromatographic experiments—is apparently sufficient to outcompete the previously revealed very strong binding of this class of corroles to isolated serum proteins [46, 47]. This was not the case for the iron complex **2-Fe**, a porphyrin with symmetrically distributed sulfonic acid head groups, which we have confirmed not to bind to any of the lipoproteins classes [8]. On the other hand, the amphipolar hemin, with its two carboxylic acids located in a non-symmetric fashion on the porphyrin ring, did bind to lipoproteins on the expense of other serum components. Together the results indicate that the charge distribution on the macrocycle is the main determinant of its association with serum components.

The very selective serum distribution pattern of **1-Fe**, compared with those of **1-Mn** and **1-Ga**, points towards distinctive binding modes/motifs of the former in its interaction with lipoproteins. As these corroles differ only in the identity of the central metal, the chelated iron(III) ion is apparently responsible for the characteristics of **1-Fe** association (vide infra). Regarding the other two corroles, similarity of spectral changes with both HDL2 and LDL points toward similar binding modes of each compound to both lipoprotein classes (Figs. 3.3, p. 19 and 3.4, p. 20). The lack of CD signals for **1-Mn**- and **1-Ga**-lipoprotein conjugates indicates that they associate to non-chiral components therein. Further support for this conclusion comes from the fact that liposomes induce the same spectral changes as do lipoproteins (Fig. 3.11, p. 24) and added amino acids do not affect the absorbance. It may hence be concluded that these two corroles are imbedded within the amphipolar phospholipid layer present in all lipoproteins. Indeed, the higher corrole percentage in HDL2 relative to LDL and in LDL relative to VLDL may be easily explained on the basis of the relative molar concentrations of these particles within serum (Table 4.1 and Fig. 4.1) [9]. Nevertheless, HDL3 molar

4.1 Interaction of Corroles with Lipoproteins

**Table 4.1** Properties of lipoproteins

| Lipoprotein type | Concentration in serum (μM) | Density (g L$^{-1}$) | Diameter (nm) | Mean Mw (Da) | Protein (% weight) |
|---|---|---|---|---|---|
| VLDL | ~0.2 | 0.940–1.006 | 30–70 | $7.5 \times 10^6$ | 5–10 |
| LDL | ~0.9 | 1.006–1.063 | 15–25 | $2.5 \times 10^6$ | 20–25 |
| HDL2 | ~3.5 | 1.063–1.115 | ~10.5 | $4.0 \times 10^5$ | 35–40 |
| HDL3 | ~3.5 | 1.115–1.210 | 7.5–10 | $2.8 \times 10^5$ | 45–65 |

concentration is at least as high as that of HDL2 [10], and yet the lowest **1-Mn** and **1-Ga** binding percentage recorded was for HDL3. This may be attributed to the very low lipid content and surface area of the smaller and denser HDL3, which are the lowest among all four lipoprotein subclasses, so that HDL3 probably cannot physically bind many corrole molecules. The importance of lipid content and surface area for corrole binding may be appreciated by examining the previously mentioned maximal corrole-binding capacity of purified lipoproteins, which in LDL was fourfold higher than in HDL (Table 3.1, p. 18).

### 4.1.2 Mode of 1-Fe Binding to Lipoproteins

The first clue for understanding the highly selective binding of the iron(III) corrole to particular lipoproteins was provided by the differences in the absorbance spectra of **1-Fe**/HDL2 versus **1-Fe**/LDL, thus pointing towards different binding modes therein (Fig. 3.2, p. 19). While the absorbance spectrum of **1-Fe**/LDL was very similar to that of **1-Fe** in buffer (the only difference was a 10 nm red-shift of the Soret band in LDL relative to buffer), the spectrum of **1-Fe**/HDL2 contained a new band at 620 nm, that was more intense than the 550 nm band (and a 20 nm red-shift relative to buffer of the Soret band). What is more, serum saturated with **1-Fe** displayed lower selectivity for HDL2 binding, with a distribution pattern more like that displayed by **1-Mn** and **1-Ga**—only 75 % bound to HDL2, 20 % to LDL and 5 % to VLDL and HDL3. The spectrum of the **1-Fe**/HDL2 conjugates in this case was not identical to that obtained under limited corrole concentrations (Fig. 3.5, p. 20). In fact, the spectrum of **1-Fe**/HDL2 from serum overloaded with corrole may be considered as a superposition of the spectra of **1-Fe**/HDL2 conjugates under low corrole concentrations and of **1-Fe**/LDL conjugates. The Soret band was 14 nm red shifted relative to buffer solutions, and while the 620 nm band was still clearly seen, it was less intense than the 550 nm band. This leads to the conclusion that HDL2 binds **1-Fe** by different binding modes (Fig. 4.2), of which the highest affinity one is specific to HDL2 (protein binding, vide infra) and the weaker one is general for lipoproteins (binding to the lipidic phase in the same manner as **1-Mn** and **1-Ga**). The second binding mode becomes apparent already at a **1-Fe**/HDL2 ratio of 2, indicative that there is only one super-strong binding site per HDL2 particle.

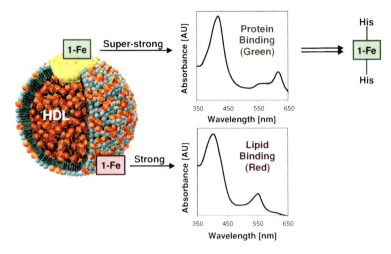

**Fig. 4.2** The two different binding modes of **1-Fe** to HDL2. **1-Fe** may bind to HDL2 by strong binding to the lipids (providing *red* solutions) or by super-strong binding to proteins (providing *green* solutions as when coordinated with two His moieties)

Confirmation of the two different binding modes hypothesis, as well as additional insight regarding the nature of binding, was obtained by CD spectroscopy. This approach was adopted because of the significant differences between HDL and LDL in terms of their lipid versus protein contents (Table 4.1). In addition, the proteins present in each of these lipoproteins are very much different. While LDL contains only one very big protein, apoB$_{100}$ (Mw = 500 kDa) [9], HDL particles possess a large variety of much smaller apolipoproteins (A's, E's, C's; Mw $\leq$40 kDa), as well as various enzymes (PON1, LCAT, and more) [10, 11]. Native LDL and HDL provide CD spectra with bands in the UV range due to their protein(s), but not in the visible. On the other hand, the visible light absorbing non-chiral corrole may display a CD spectrum only when located within a chiral environment. A very rich CD spectrum in the visible range was obtained for **1-Fe**/HDL2 solution, while no visible-range CD signals were displayed by **1-Fe**/LDL solutions (Fig. 3.8, p. 23), clearly pointing towards specific interactions of the corrole with proteins of HDL2, but not with apoB$_{100}$ of LDL. The lack of CD signals for **1-Fe**/LDL, as well as the analogous absorbance spectra observed for **1-Fe** with LDL and with liposomes, further indicates that corroles bind to the lipid phase of LDL.

Independent evidence for **1-Fe** binding to HDL-associated protein-derived moieties was obtained by examining the effect of basic amino acid esters, which may coordinate to the metal center of the corrole, on the electronic spectrum of **1-Fe** in buffer. These experiments revealed that histidine, with its strongly coordinating imidazole side chain, induced spectral changes similar to those seen with HDL2 (Fig. 3.9, p. 24), while all other amino acids examined (Met, Tyr, Cys, Arg

4.1 Interaction of Corroles with Lipoproteins

and Lys) did not affect the spectrum. Titration of **1-Fe** with a histidine ester displayed isosbestic points only at low excess of the latter, consistent with the presence of three absorbing species at *higher* histidine excess: **1-Fe** coordinated to water only, to one histidine, and to two histidine molecules. The absorbance spectrum (especially concerning the 620 nm band) was similar to that of **1-Fe**/HDL2 conjugates only under conditions favoring the formation of bis-histidine-coordinated **1-Fe**. Accordingly, it may be concluded that **1-Fe** binds to HDL2 by interaction of the metal center with two closely positioned histidine residues from either one or two HDL2-associated apolipoproteins/enzymes. This may explain the high binding affinity and selectivity of **1-Fe** to HDL2 on the expense of all other serum components. The much lower affinity of **1-Mn** for histidine binding, demonstrated by the much higher histidine excess needed to induce spectral changes, may explain the lack of HDL2-selectivity for this corrole.

The quite low affinity of **1-Fe** for bis-histidine coordination in the homogenous buffer solution versus the high one in HDL2 is quite illuminating. It clearly suggests supramolecular arrangement of proteins in HDL2 that is perfectly suited for the interaction, a situation that apparently does not exist in HDL3. The excellent catalytic activity of HDL-conjugated corrole (vide infra), despite of the dominancy of the hexa-coordination of the metal therein, may be explained by fast equilibrium between the mono- and bis-histidine coordinated **1-Fe**. A more detailed explanation for the discrimination between HDL2 and HDL3 by **1-Fe** requires the identification of the specific protein(s) involved in the binding of the corrole, which is quite a formidable task because HDL contains numerous proteins and enzymes. The effect of two proteins on **1-Fe** binding was nevertheless examined because of their quite different abundance in HDL subclasses. Liposomes containing either apoE (resides mostly in HDL2) [12] or PON1 (found mainly in HDL3) [13] did however not induce the formation of the 620 nm band that is characteristic of **1-Fe**/HDL2 conjugates (Fig. 3.10, p. 24). Furthermore, the 620 nm band did not disappear in serum of knockout mice for neither of these proteins (Fig. 3.12, p. 26). Together these results suggest that neither apoE nor PON1 are responsible for **1-Fe** binding to HDL. Another possibility is that PON1 may disturb **1-Fe** binding to HDL3, and this was examined by incubating HDL2 with PON1 prior to corrole addition. Indeed, this practice resulted in an absorbance spectrum more similar to that obtained for **1-Fe**/LDL conjugates (Fig. 3.13, p. 26), a less red-shifted Soret band and a larger intensity of the 550 relative to the 620 nm bands, indicating a shift from protein to lipid binding. PON1 has been shown to interact with HDL through apoAI binding [14], suggesting that binding of **1-Fe** to HDL was through apoAI. However, experiments with serum from apoAI knockout mice dismissed this hypothesis, as the 620 nm peak was still seen in the absence of apoAI (Fig. 3.12, p. 26). Further attempts to elucidate the protein(s) assembly responsible for **1-Fe** binding to HDL were postponed to future investigations.

## 4.2 The Effect of Corroles on Oxidative-Stress Induced Lipoprotein Damage

### 4.2.1 Lipoprotein Oxidation Products

Lipoproteins contain many substrates that may be oxidized—fatty acids, proteins and cholesterol—and a variety of methods exist for determining the amount of lipoprotein oxidation [9, 15]. Polyunsaturated fatty acids are the initial oxidation targets of LDL (Scheme 4.1), as they contain bis-allylic hydrogen atoms that may easily be abstracted by an oxidant. The resulting carbon-centered radical rapidly reacts with dioxygen, leading to double bond conjugation and production of a lipid peroxide. This peroxide may then abstract a bis-allylic hydrogen from a neighboring fatty acid to form a hydroperoxide while propagating the radical chain reaction. In the presence of transition metals hydroperoxides do not accumulate as they are rapidly cleaved to form aldehydes.

Similar to the majority of studies, our early investigations focused on LDL oxidation, as oxLDL is the main contributor to the development of atherosclerosis [16–18]. However, the elucidation of highly preferred corrole binding to serum HDL, together with the lately discovered major role of dysfunctional HDL in the atherosclerotic process [10, 19], shifted our attention to HDL oxidation and its consequences. Since lipid damage to HDL is much less pronounced than in the case of LDL (due to the lower lipid content of the former, 55 vs. 75 %, Table 4.1), attention was given in this case also to protein oxidation.

Initiation of lipoprotein oxidation by copper ions is considered a facile route for obtaining in vitro oxidized lipoproteins with features similar to those of in vivo sources [20]. The copper(II) ions added to the lipoproteins are first reduced by hydroperoxides and other reducing agents naturally occurring in lipoproteins [21], and the thus formed copper(I) may react with dioxygen to generate $O_2^-$ and consequently $H_2O_2$ [22]. The metallocorroles were shown to catalytically decompose these two species [23, 24], and so the effect of corroles on copper ions induced lipoprotein oxidation was examined by measuring the formation of hydroperoxides and aldehydes after a few hours from initiation. Both measurements indicated **1-Fe** as an inhibitor of lipoprotein oxidation from a concentration of $\geq 2.5$ μM, almost completely preventing the formation of the two oxidation products. However, contradicting results were displayed by **1-Mn**, as it reduced formation of hydroperoxides but largely increased formation of aldehydes. The non-redox active **1-Ga**, which does not serve as an ROS/RNS decomposition catalyst, expectedly showed no effect on lipoprotein oxidation under all examined concentrations (Fig. 3.14, p. 26). To gain deeper understanding, the kinetics of copper ions induced lipoprotein oxidation at the presence of 2.5 μM of **1-Fe** or **1-Mn** was examined (Fig. 3.15, p. 27). In the absence of corrole, a delay time was seen between the addition of oxidant and the formation of oxidation product. This was due to dietary (sacrificial) antioxidants naturally present within the lipoproteins, that are the first to be modified by the ROS/RNS [9]. This delay time is longer for LDL relative to

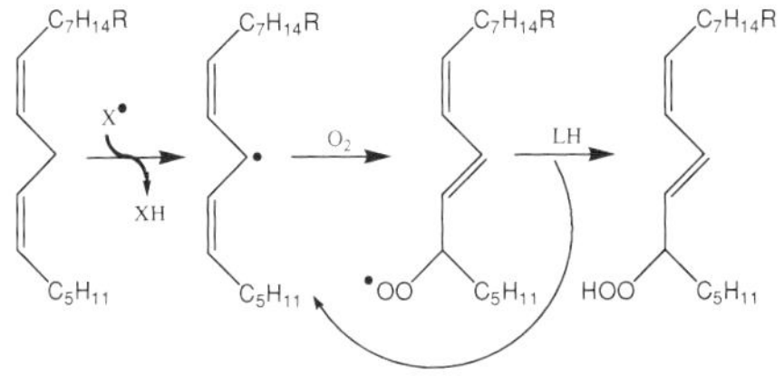

## 4.2 The Effect of Corroles on Oxidative-Stress Induced Lipoprotein Damage

**Scheme 4.1** The process of lipid peroxidation. Lipid peroxidation of linoleic acid, the most abundant poly-unsaturated fatty acid in LDL. Only one possible hydroperoxide product is shown

HDL, as the former contains a higher amount of such intrinsic antioxidants (an average of 10 vs. <1, respectively) [25, 26]. Enrichment of lipoproteins with sacrificial antioxidants elongates the oxidation delay time, but only by several minutes per each antioxidant molecule. Pre-treatment with the corroles showed again that **1-Fe** was an excellent catalytic antioxidant, completely avoiding the formation of lipid oxidation products (even after 24 h from initiation). The same kind of experiments revealed that **1-Mn** is actually a pro-oxidant, since it shortened the time for initiation and increased the amount of oxidation. The kinetic examination thus settled the conflict regarding the action of **1-Mn**: the lower amount of hydroperoxides recorded in the presence of this compound in the dose-dependence experiments was because measurements were performed after their maximal levels were reached, and they were already transformed into aldehydes.

Peroxynitrite is a RNS considered to be involved in the development of many diseases, including atherosclerosis [4, 9]. As the corroles have been shown to efficiently detoxify peroxynitrite in a catalytic fashion [27], their ability to inhibit peroxynitrite induced lipoprotein oxidation was examined. The half-life of peroxynitrite under physiological relevant conditions is only about 1 s [28], and so very large quantities of the oxidant need to be added in vitro to get significant lipoprotein damage. This is considered not a good model for in vivo occurring damage, as peroxynitrite forms in the body in low but continuous fluxes. Accordingly, oxidation was triggered by the addition of SIN-1, a reagents that slowly and spontaneously decomposes at neutral pH to produce $O_2^-$ and NO, consequently forming peroxynitrite in a fashion that mimics its in vivo formation [29]. As we found no routine protocol for SIN-1 induced lipoprotein oxidation, the first step was to determine the appropriate working concentration. All lipid and protein oxidation parameters measured (conjugated dienes, hydroperoxides, tryptophan fluorescence, and electrophoretic pattern), were largely increased with increasing SIN-1 concentrations, and started to level off at about 500 μM (Figs. 3.16, p. 28 and 3.17, p. 28). An exception was the amount of aldehydes, which remained very low under all SIN-1 concentrations, as transition metal ions are needed to catalyze hydroperoxides breakdown to aldehydes. Protein tyrosine nitration is considered a fingerprint of peroxynitrite, and indeed increasing SIN-1 concentrations elevated the amount of nitro-tyrosine; however meaningful increase was seen only at 500 μM or higher (Fig. 3.18, p. 29). ApoAI nitration was observed only from 500 μM of SIN-1, while

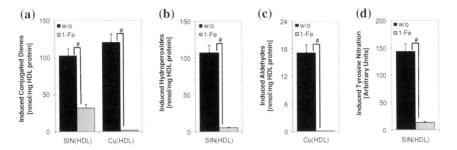

**Fig. 4.3** The effect of **1-Fe** on HDL oxidative damage by SIN-1 or CuSO$_4$. The effect of 50 μM of **1-Fe** on overnight oxidation of HDL with SIN-1 or copper ions: conjugated dienes (**a**), hydroperoxides (**b**), aldehydes (**c**), and tyrosine nitration (**d**). $^{#}p < 0.005$, $^{##}p < 0.05$ (vs. w/o corrole)

1,000 μM gave a very high amount of protein nitration, not allowing for identification of specific protein bands. Taken together, 500 μM of SIN-1 was selected as an appropriate concentration to induce significant but reasonable lipoprotein damage. Pre-treatment of lipoproteins with **1-Fe** provided the former with protection against oxidation from a 50 μM concentration[2]; and nitration was efficiently prevented already at 10 μM of **1-Fe**. **1-Mn** not only increased oxidative damage, but also highly elevated protein nitration.

The effects of 50 μM of **1-Fe** on both SIN-1 and copper ions-induced oxidation of HDL are summarized in Fig. 4.3. For treatment with SIN-1, **1-Fe** brought about a decrease of 70 and 95 % in conjugated dienes and hydroperoxide formation, respectively, and a 90 % reduction in tyrosine nitration. For copper-induced oxidation, **1-Fe** completely eliminated the formation of both conjugated dienes and aldehydes.

### 4.2.2 Lipoproteins Biological Activities

It is well known that oxidative modifications of HDL and LDL turns the former less anti-atherogenic and the latter more pro-atherogenic than their native counterparts [30, 31]. As **1-Fe** is highly beneficial for preventing formation of lipoprotein oxidation products and as it binds in serum with high preference to HDL, its ability to prevent oxidative stress induced HDL loss of functions was examined (Figs. 3.21 on p. 31 and 4.4).

The cholesterol efflux capacity of HDL was somewhat reduced by oxidation, an effect that was alleviated when **1-Fe** was added to HDL prior to the oxidant.

---

[2] The effective concentration of **1-Fe** for the SIN-1 induced oxidation was tenfold higher than in the copper ions induced oxidation. However, lipoprotein concentration was also increased tenfold in the former relative to the latter. See experimental section for further details.

4.2 The Effect of Corroles on Oxidative-Stress Induced Lipoprotein Damage

| % Cholesterol Efflux | |
|---|---|
| HDL | 37.5 |
| oxHDL | 30-32 |
| 1-Fe/HDL | 34-36 |

| % Oxidative Stress Inhibition | |
|---|---|
| HDL | 60 |
| oxHDL | 25-30 |
| 1-Fe/HDL | 85 |

| % Apoptosis Inhibition | |
|---|---|
| HDL | 75 |
| oxHDL | 15-30 |
| 1-Fe/HDL | 70-90 |

**Fig. 4.4** The effects of oxidation and of **1-Fe** on HDL activities. The cholesterol efflux, oxidative stress inhibition and apoptosis inhibition abilities of native HDL, oxHDL and **1-Fe**/HDL conjugates

The anti-apoptotic activity of HDL was greatly reduced by oxidative damage, from 75 % inhibition of apoptosis for native HDL to only 15–30 % for oxHDL, but pretreatment with the corrole provided full protection against this function impairment. Most impressive was the effect of **1-Fe** on the anti-oxidative capacity of HDL—the corrole not only prevented the reduction of activity, but even improved the functionality relative to native HDL. Native HDL displayed 60 % inhibition of oxLDL-induced cellular oxidative stress, oxHDL showed only 25–30 % inhibition, and **1-Fe**/HDL conjugates were highly efficient displaying 85 % inhibition. A plausible explanation for this phenomenon is that **1-Fe** present within the **1-Fe**/HDL conjugate contributes to the anti-oxidative activity by directly eliminates the *secondary* oxidants formed by the cells following oxLDL treatment. Support of this hypothesis was provided by examination of the direct effect of **1-Fe** on oxLDL. When oxLDL was incubated with **1-Fe** prior to its addition to the cells, its ability to induce oxidative stress was lower than for non-treated oxLDL (Fig. 3.22, p. 32). An alternative role for **1-Fe** in reducing the pro-oxidativity of oxLDL may be catalysis of hydroperoxides decomposition, a property that was previously demonstrated for porphyrins [32], leading to a reduced oxidative status of the oxLDL.

### 4.2.3 Mechanistic Analysis

The anti- versus pro-oxidant activity of **1-Fe** versus **1-Mn** may be rationalized based on their different catalytic mechanisms, as specifically depicted for peroxynitrite decomposition (Fig. 4.5). **1-Mn** acts by a disproportionation mechanism, forming an (oxo)manganese(V) intermediate via its reaction with one HOONO molecule. This intermediate is a strong oxidizing agent that reacts with a second HOONO molecule as to complete the catalytic cycle. However, in a biological surrounding the (oxo)manganese(V) intermediate apparently reacts with the highly available biomolecules (rather than with HOONO), thus acting as a pro-oxidant. Unlike porphyrins, the (oxo)manganese(V) corrole intermediate, with an oxidation potential of 0.6–0.7 V (vs. Ag/AgCl) [33], is not a redox couple of the nitrite formed in this first step of the catalytic cycle. Accordingly, $NO_2$ radical is not formed, and the increased nitration displayed in the presence of **1-Mn** seems to be

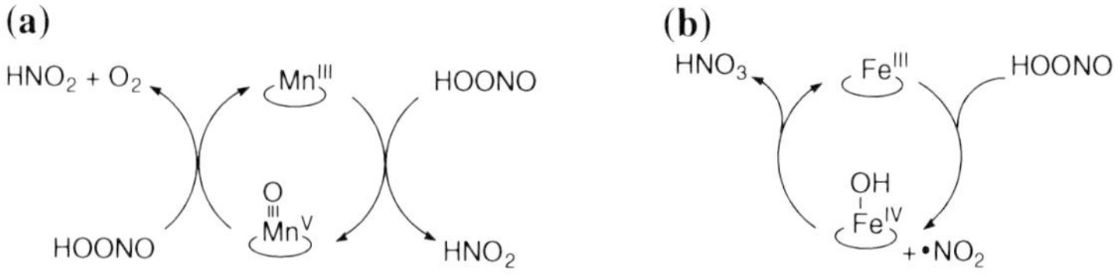

**Fig. 4.5** The catalytic mechanism for HOONO decomposition by **1-Fe** and **1-Mn**. The suggested decomposition mechanism of HOONO by disproportionation (**a**) and isomerization (**b**) for **1-Mn** and **1-Fe**, respectively

an indirect effect that results from its pro-oxidant activity (by reaction of the oxidized biomolecule with the nitrite). While $NO_2$ radical is however formed in the catalytic cycle of HOONO decomposition by **1-Fe**, the very high catalytic rates of this catalyst imply that the radical rapidly reacts with the (hydroxyl)iron(IV) intermediate, before the two diffuse from one another and cause any damage. This prevents any oxidizing and nitrating agents from being released to the bulk and harming the biological molecules.

In conclusion, the potent catalytic properties of **1-Fe** are preserved upon its conjugation to HDL, despite of the hexa-coordination state of the iron(III) atom. The conjugated **1-Fe** serves as a shield for HDL against ROS/RNS, preventing oxidative stress induced loss of HDL anti-atherogenic functions, and even improving its anti-oxidative capability relative to native HDL. **1-Fe** also reduced the pro-oxidative activity of already oxidized LDL. On the other hand, **1-Mn** served as a pro-oxidant and the non-redox active **1-Ga** was inert regarding lipoprotein oxidation.

## 4.3 Interactions of Corroles with Macrophages

When examining cellular effects of an applied compound it is important to know its level of accumulation within the cells. Intracellular therapeutic agents are commonly identified by fluorescence measurements of an appropriate drug-dye conjugate. This approach is problematic, as quite often the dye is larger than the drug itself, and hence may affect the interaction of the drug with the cells. In the case of transition-metal complexes this practice may not be exploited at all due to fluorescence quenching by the metal. In the case of corroles, these two problems may be solved by using an analogous non-transition metal complex, which results in a fluorescent compound almost identical to the investigated agent. Accordingly, the fluorescent **1-Ga** was used for estimating cellular accumulation of amphipolar corroles in all cases reported so far [16, 19], and was the starting point in the current study as well. Fluorescence microscopy follow up of **1-Ga** revealed to two important observations: the compound did not break down upon its entrance to cells (as corrole fragments are not fluorescent), and it resided in the cytoplasm and not in the cell nucleus. However, using **1-Ga** for characterizing cell accumulation of **1-Fe** suffers from several drawbacks. First, **1-Ga** is commonly used in cancer models for killing cells rather than protecting them [34]; and indeed macrophage survival was found to be heavily reduced following incubation with **1-Ga** (Fig. 3.26, p. 34). Second, cellular trafficking of negatively charged molecules is highly dependent on the molecules interaction with serum components (Sect. 4.3.2, p. 53); and we found that serum interactions of **1-Fe** and **1-Ga** are different (Sect. 4.1.1, p. 45). This called for the development of a direct method for measuring **1-Fe** at nanomolar concentrations.

## 4.3.1 Chemiluminescence-Based Detection of 1-Fe Accumulation in Macrophages

Two methods have been introduced for quantification of transition metal porphyrins at sub-micromolar concentrations: measurement of iron(III) porphyrin by atomic absorption and ex vivo replacement of the porphyrin-chelated manganese(III) by zinc(II) and quantification of the zinc(II) porphyrin by fluorescence [35, 36]. The first method requires a very large quantity of porphyrin-containing cells, which in many cases is not practical. The second method is non-trivial and limited to cases in which the metal may be substituted [not the case for iron(III) complexes of both porphyrins and corroles] [37]. We now introduce a new method for the direct quantification of iron complexes of corroles and porphyrins at nanomolar concentrations, based on a catalytic rather than photophysical property. In biochemical research, utilization of catalytic properties of enzymes for detection purposes is a routinely employed method, serving as the basis of commonly used applications such as Western Blot analyses and ELISA assays. The most frequently used enzyme for such detection reactions is HRP, which contains an iron-chelated porphyrin as its prosthetic group. HRP catalyzes the oxidation of luminol, a reaction that may also be catalyzed by free iron(III) porphyrin complexes [38], leading to the emission of blue light, that serves as the detection reaction. This reaction is also used for forensic investigations to detect blood traces in a crime scene, as the iron(III) porphyrin-containing hemoglobin present in the blood also catalyzes the luminal chemiluminscent reaction.

**1-Fe** was highly potent for catalyzing luminol oxidation even at concentrations as low as 10 nM in cell-free systems, while **1-Mn** was relatively inefficient even at a concentration of 100 nM. The **1-Fe**-catalyzed reaction was most efficient at basic pH, with initial luminescence intensity reducing twofolds when changing the pH from 11 to 9, and another tenfold decrease in neutral pH. The reaction profile also depended on the relative amounts of luminol and $H_2O_2$: when using a tenfold excess of $H_2O_2$ relative to luminol the initial luminescence intensity was very high but the kinetics was very fast, whereas equimolar concentrations of the two gave lower emission but much more easy to follow kinetics. Applying equimolar concentrations of $H_2O_2$ and luminol provided very low emission in the presence of cell debris, but

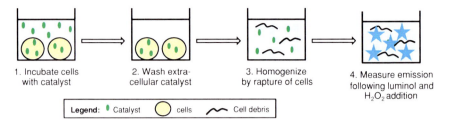

**Fig. 4.6** The experimental steps for measuring intracellular **1-Fe** and iron(III) porphyrins. The chemiluminescent method for determining intracellular amount of **1-Fe**, **2-Fe**, **3-Fe**, and hemin

excess $H_2O_2$ displayed kinetic curves that could be integrated to give reproducible results. Integration of the kinetic curves resulted in a linear calibration curve, allowing for the quantification of cellular-derived **1-Fe** down to a concentration of 30 nM (Fig. 3.31, p. 36), much below the micromolar concentrations needed to detect this compound by absorption measurement. This discovery was used for the appreciation of intracellular levels of the corrole by incubation of **1-Fe** with macrophages, removing the non-internalized fraction, rupture of cells, and measurement of emission kinetics following addition of luminol and $H_2O_2$ to the macrocycle-containing cell debris (Fig. 4.6). Using the new method, **1-Fe** uptake and clearance from J774.A1 macrophage cells was followed (Fig. 3.32, p. 38). Interestingly, the presence of **1-Fe** caused elevation of cellular viability, an issue that is discussed in the following section.

The level of intracellular **1-Fe** was increased as the amount of the compound added to the culture medium was raised, approaching saturation at about 100 µM of applied corrole (Fig. 3.33, p. 38). Under these conditions only 1 % of the corrole added to the medium was found intracellularly, indicating that **1-Fe** uptake proceeds via a specific mechanism that reaches saturation even when excess corrole is still present in the medium. It is customary to normalize results obtained from experiment with cells relative to the amount of cell protein levels, which are assumed to be proportional to the amount of cells. Unfortunately, determination of protein concentrations could not be conducted after the chemiluminescent measurement, and so the number of cells in each well was evaluated based on previous experience with the J774.A1 cell line to be $10^5$ cells well$^{-1}$. Accordingly, the amount of intracellular **1-Fe** under saturation was 100 pmol $10^{-5}$ cells, which may be translated to 1 fmol cell$^{-1}$ or $6 \times 10^8$ molecules cell$^{-1}$ (or 250 µM assuming a cell volume of $4 \times 10^{-9}$ mL). This is a meaningful number, as it is 2–3 orders of magnitude larger than the cellular concentration of the most important and extremely fast acting ($k_{cat} = 2 \times 10^9$ $M^{-1}$ $s^{-1}$) enzymatic antioxidant, SOD: $2.5 \times 10^5$ molecules cell$^{-1}$ in erythrocytes and $1 \times 10^6$ molecules cell$^{-1}$ in lymphocytes (translated from 1.4 and 5.6 ng cell$^{-1}$, respectively) [39]. It is also only one order of magnitude lower than the about 15 fmol cell$^{-1}$ of glutathione, the major (but slowly acting and non-catalytic) antioxidant present in cells, shown to reside within macrophages [40, 41].

## 4.3.2 Comparison of Accumulation in Macrophages of 1-Fe and Related Iron(III) Porphyrins

The new chemiluminescence detection method was also applied for three iron(III) porphyrin complexes (Scheme 1.3): two synthetic symmetric porphyrins, one with negative and one with positive charges (**2-Fe** and **3-Fe**, respectively), and the natural porphyrin hemin. **2-Fe** was an extremely efficient catalyst, displaying very strong emission even at 7 nM. **3-Fe** and hemin were moderately efficient, providing calibration curves with a slope that was an order of magnitude lower than that for **2-Fe**. **1-Fe** was the poorest catalyst for luminol oxidation, an observation consistent with its unmatched potency (relative to other synthetic antioxidants) regarding the catalytic decomposition of hydrogen peroxide [23], the reaction that competes with the oxidation of luminol. Using the new detection method, the cellular accumulation of the corrole and the three porphyrins could be directly compared (Table 4.2; Fig. 3.35, p. 39). **1-Fe**, **2-Fe** and hemin are all negatively charged and may hence be predicted not to be able to cross the cell membrane spontaneously. They nevertheless did accumulate within cells, but the intracellular concentration of **2-Fe** was $\geq 5$ times smaller than that of **1-Fe**, 8 versus 40 pmol cell$^{-1}$. This may be attributed to structural differences between the two compounds which lead to differences in their interactions with potentially cell-internalizing serum proteins. **2-Fe** has four charges distributed symmetrically relative to its macrocyclic skeleton, while the two charged groups of **1-Fe** are located on only one side of the molecule, turning it amphipolar (Scheme 1.3, p. 8). Indeed, **1-Fe** (as well as amphipolar hemin) conjugated spontaneously and extremely strong to serum lipoproteins, while this was not the case for **2-Fe** (Sect. 3.1.2.2, p. 18) [42]. Taken together, the significantly larger intracellular concentrations of **1-Fe** relative to **2-Fe** clearly point towards serum proteins as cell internalization vehicles for the negatively charged complexes.

Based on the above, intracellular accumulation of hemin was expected to be comparable to that of **1-Fe**, but turned out to be much higher (140 pmol cell$^{-1}$). The intracellular concentration displayed by **3-Fe** (50 pmol cell$^{-1}$) was however not surprising and well in the range previously reported for iron(III) complexes of pyridinium-substituted porphyrin derivatives [35, 43], as positively charged compounds may easily cross the cell membranes. However, parallel analysis of cell viability (Table 4.2; Fig. 3.35, p. 39) showed that the results obtained for **3-Fe** and hemin were misleading, as these compounds significantly reduce cell survival to as low as 45–60 % relative to control cells, a process that is accompanied by increased membrane permeability. These findings are actually consistent with hemin being a well established pro-oxidant [44], and with the DNA-cleaving ability [45, 46] associated to the quite efficient antioxidant **3-Fe** [47, 48]. On the other hand, **1-Fe** increased cell viability up to 145 %, which could reflect the effect of this catalytic antioxidant on the attenuation of damage induced by the quite significant oxidative stress that is naturally present in cell-cultured macrophages, whereas **2-Fe** had no significant effect on cell survival. This issue is further discussed in the following section.

**Table 4.2** Cellular accumulation of **1-Fe**, **2-Fe**, **3-Fe** and hemin and consequential cell survival

|  | 1-Fe | 2-Fe | 3-Fe | Hemin |
|---|---|---|---|---|
| Cellular concentration (pmol $10^{-5}$ cells) | 40 | 8 | 50 | 140 |
| Cellular viability (% of control) | 145 | 115 | 45 | 60 |

## 4.4 The Effect of Corroles on Macrophage Atherogenicity

### 4.4.1 Oxidative-Stress Induced Damage

Macrophages are main contributors to increased oxidative burden in the artery wall, and as their oxidative status increases so does their ability to mediate lipoprotein oxidation [49]. It is hence important to prevent arterial macrophage oxidation. **1-Fe** and **2-Fe**, the two complexes that did not reduce basal cell viability in macrophages, were hence examined for their capability to protect the cells against oxidative stress. This was done by deliberately exposing cells loaded with these compounds to various agents that initiate oxidative stress-induced cell death. The toxins used for that purpose may be divided to two groups: (a) primary oxidants, including $H_2O_2$ and peroxynitrite (progressively formed from the applied SIN-1) [29], which the antioxidants may directly neutralize in a catalytic fashion; and (b) indirect oxidants, including oxLDL [50] and a combination of LPS and INF-$\gamma$ [51–53], which cause immuno-activation of the cells for the production of ROS. NADPH oxidase is involved in both methods of cellular activation, which leads to the formation of a high flux of $O_2^-$ (that may dismutate to $H_2O_2$). In the latter system nitric oxide synthase (NOS) is also activated, leading to production of NO, which together with the $O_2^-$ may form peroxynitrite.

The results revealed that **1-Fe** was beneficial for preventing cell death in all the applied systems, with a twofold increase in cell viability in most cases, while **2-Fe** had no significant effect on cell survival relative to cells not treated with any catalytic antioxidant (Table 4.3; Figs. 3.36 and 3.37, p. 41). **1-Fe** dismutates $O_2^-$ and neutralizes $H_2O_2$ extremely rapidly [23], thus avoiding the formation of any toxic species by this route, explaining its efficiency in systems which these toxins are involved (addition of $H_2O_2$ or oxLDL). Even in cases that $O_2^-$ may still react with nitric oxide to form peroxynitrite (addition of SIN-1 or LPS + INF-$\gamma$), the latter may also be efficiently detoxified by **1-Fe** [27]. Of particular relevance is the effectiveness of **1-Fe** against oxidants formed *in cellulo* following cellular activation, since this confirms that the iron corrole identified by the chemiluminescence method (Sects. 4.3.1 p. 55 and 4.3.2, p. 57) was in fact intracellular and not just cell-associated. The superiority of **1-Fe** relative to **2-Fe** is in line with both its higher intracellular concentration and the larger catalytic rates displayed by it for decomposition of all major ROS/RNS (Table 4.4).

While many metalloporphyrins have been shown to be very efficient catalysts for dismutation of $O_2^-$ [55], most are very poor catalysts for $H_2O_2$ decomposition due to extensive catalyst bleaching [56]. Selected porphyrins have been shown

## 4.4 The Effect of Corroles on Macrophage Atherogenicity

**Table 4.3** The effect of **1-Fe** and **2-Fe** against oxidative-stress induced cellular death

|      | $H_2O_2$ | SIN-1 | oxLDL | LPS + INF-$\gamma$ |
|------|----------|-------|-------|--------------------|
| w/o  | 20       | 28    | 56    | 40                 |
| **1-Fe** | 54   | 48    | 88    | 80                 |
| **2-Fe** | 21   | 27    | 58    | 47                 |

The numbers represent % cell survival (relative to control) following exposure of cells loaded with the iron complexes to various toxins

**Table 4.4** Catalytic rates for ROS/RNS decomposition and cellular concentrations of **1-Fe** and **2-Fe**

| Compound | SOD activity $(M^{-1} s^{-1})$ | CAT activity $(M^{-1} s^{-1})$ | ONOOH decomposition $(M^{-1} s^{-1})$ | Cellular concentration[a] (fmol cell$^{-1}$) |
|----------|-------------------------------|-------------------------------|---------------------------------------|----------------------------------------------|
| **1-Fe** | $3 \times 10^6$ [24]          | 6,400 [23]                    | $3 \times 10^6$ [27]                  | 0.4                                          |
| **2-Fe** | $6 \times 10^5$ [48]          | None [54]                     | $9 \times 10^5$ [47]                  | 0.08                                         |

[a] Concentration following 24 h incubation with 20 µM of the compound

beneficial against both $O_2^-$ and $H_2O_2$ toxicity to cultured cell [57–59], but other metalloporphyrins have in fact been employed for inducing cell death by transforming the less toxic $O_2^-$ to the more harmful $H_2O_2$ [35, 60]. Metalloporphyrins also serve well for eliminating peroxynitrite [true for iron(III) complexes [61] and, in the presence of reducing agents, for manganese(III) complexes as well] [62], and they have been employed for salvage of cells both from authentic and from immuno-stimulated peroxynitrite. The investigated complexes were highly potent against bolus addition of peroxynitrite (though in most cases a minimal concentration of 100 µM metalloporphyrin was needed), but only very limited activity was demonstrated against LPS + INF$\gamma$ cellular activation [51, 63–67]. Most importantly, the vast majority of cell culture investigations was performed by adding the ROS/RNS decomposition catalysts directly to the medium, i.e., without distinguishing between extracellular and intracellular effects as done in the current study. This may be responsible for the high efficiency of the compounds against oxidants added to the culture media relative to their very low efficiency against oxidants formed *in cellulo*, with the latter case reflecting their low intracellular concentrations.

**1-Mn** was as efficient as **1-Fe** for prevention of oxidative-stress induced cellular death, which was surprising in light of the pro-oxidative character this compound displayed regarding lipoproteins (Sect. 4.2.3, p. 53). However, one must consider that cells and purified lipoproteins pose very different environments to the therapeutic compound: while lipoprotein-associated antioxidants are rapidly consumed, cellular antioxidants may be recycled, e.g. via the glutathione/glutathione peroxidase system. In a cellular environment the manganese(V) oxo intermediate formed during catalysis may hence readily react with the cellular antioxidants, thus preventing the reaction of the intermediate with vital biomolecules and recycling **1-Mn** for further ROS/RNS elimination.

## 4.4.2 Effect on Cholesterol Metabolism

Cholesterol accumulation in macrophages turns them into foam cells, the first step in the development of atherosclerotic plaques, and so the effects of **1-Fe** on the three pathways governing cholesterol homeostasis were measured. While intracellular **1-Fe** did not affect cholesterol uptake or removal, it significantly reduced de novo cholesterol biosynthesis by the cells. The percent of inhibition increased up to 35 % in correlation with the enhancement in the intracellular level of the corrole up to 30 pmol cell$^{-1}$ (Fig. 3.39, p. 42). Reduced cholesterol biosynthesis was also displayed in vivo by macrophage cells harvested from the peritoneum of mice treated with **1-Fe** relative to cells from non-corrole-treated mice, which was 50–60 % for young mice and 30 % for aged mice (Fig. 3.40, p. 42).

Preliminary investigations for gaining clues about the mechanism of action of **1-Fe** showed that this compound does not affect cholesterol biosynthesis when supplying the cells with mevalonate instead of acetate, a downstream substrate. This clearly suggests that **1-Fe** affects the biosynthetic pathway at its early stages, similar to statins. However, the structural differences between **1-Fe** and statins argue against an identical mechanism of competitive inhibition of HMGCR [68], the rate determining step of cholesterol biosynthesis; and additional investigations displayed that **1-Fe** also does not affect this enzyme's mRNA expression, protein expression, and protein phosphorylation level. A plausible explanation for the effect displayed by **1-Fe** is that the redox-active complex interferes with the NADPH-dependant process catalyzes by HMGCR (Scheme 4.2). This was tested in a cell-free system, an experiment that indicated that **1-Fe** actually catalyzed rather than inhibited the reaction. However, the applied assay followed only the disappearance of NADPH and not the formation of mevalonate, and thus the results may be misleading: **1-Fe** may react directly with NADPH thus depleting the pool available for the HMGCR catalyzed reaction. More in-depth investigations are needed for understanding the effect of **1-Fe** on cellular cholesterol biosynthesis.

The last point to emphasize is that under identical conditions, **1-Fe** reduced cholesterol biosynthesis more efficiently than pravastatin and fluvastatin (35, 10, and 25 % respectively), with a combined effect of the corrole with either statin higher than the effect of each component alone (Fig. 3.39, p. 42). The lower efficiency displayed here by the statins relative to publications in the literature [69, 70] results from the different procedure applied in this study, in which the cells were incubated overnight in the absence of the examined agent, and only then assayed for cholesterol biosynthesis. The outcome of this procedure was that the actual concentrations of the therapeutic compounds at the assay time were much lower than what was originally added, as specifically showen for **1-Fe** to be only 1 %. The option of using **1-Fe**/statin combination therapy is particularly appealing, as it may allow for lower statin doses and thus to less side effects [71–74]. In contrast to our results with the catalytic antioxidant **1-Fe**, natural antioxidants have been shown to interfere with the activity of statins [75, 76].

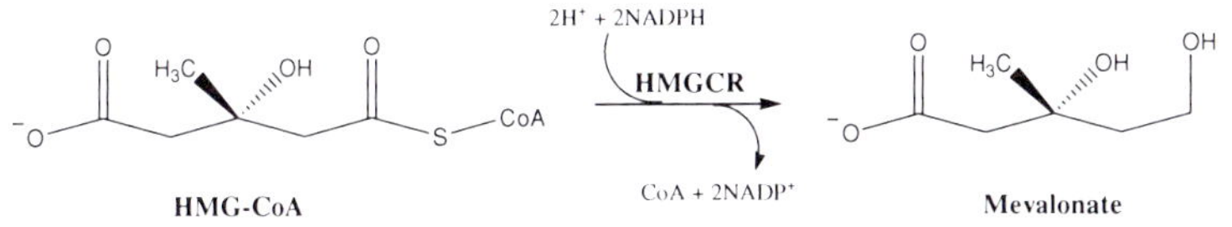

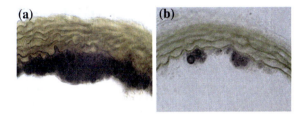

**Scheme 4.2** The rate determining step of cholesterol biosynthesis. The enzyme HMGCR reduces HMG-CoA to mevalonate by using two equivalents of NADPH

**Fig. 4.7** Atherosclerotic lesions from control and **1-Fe**-treated mice. Photos of the aortic arch from control (**a**) and **1-Fe**-treated (**b**) young mice. The lipidic matter is stained with a *black* dye

## 4.5 The Effect of Corroles on Atherosclerosis Development in Mice

The in vivo effects of corroles were examined by their oral administration to $E^0$ mice, that are genetically engendered to develop atherosclerosis. 12-week old mice were treated with **1-Fe**, **1-Mn** or **1-Ga** for 12 weeks at a dose of 10 mg kg$^{-1}$ day$^{-1}$ (Fig. 3.44, p. 44). **1-Ga** was used as a negative control, as it is not a catalytic antioxidant, and hence not expected to affect atherosclerosis development; and indeed it did not inhibit lesion growth. Although **1-Mn** served as a pro-oxidant toward isolated lipoproteins, its in vivo activity was somewhat positive (16 % reduction in lesion area) as was also displayed in cell cultures. **1-Fe** was highly beneficial for reducing plaque formation (60 % reduction in lesion area) in accordance with its in vitro potent activities. This may be easily appreciated from lipid staining photos of the aortic arch of control and treated mice (Fig. 4.7). In a second mice experiment a high-risk high-gain approach was adopted, selecting for the experiment mice that were already 30 weeks old, an age at which signs of lesion development are already evident and lesion growth continues rapidly [77]. This experiment differs significantly from the previous one, in which the compounds were supplied to asymptomatic young mice (according to well accepted procedures) [78]. We believe this model represents a more plausible potential patient, which will start drug treatment only after ample evidence that he is indeed developing CVD. The aged mice were treated for 12 weeks at a dose of 10 mg kg$^{-1}$ day$^{-1}$ within their drinking water; and **1-Fe** still reduced plaque formation to a significant extent (30 % reduction in lesion area). A comparison with similar experiment performed with pomegranate juice (PJ), one of the most potent dietary antioxidants, may serve to illustrate the novelty of the iron corrole regarding attenuation of lesion formation: 44 % reduction for young mice treated with PJ [79] relative to 60 % with **1-Fe**, and only 17 % decrease for aged mice treated with PJ [80] relative to 30 % with **1-Fe**.

# References

1. Liang, L.-P., Huang, J., Fulton, R., Day, B.J., Patel, M.: An orally active catalytic metalloporphyrin protects against 1-methyl-4-phenyl-1,2,3,6-tetrahydropyridine neurotoxicity in vivo. J. Neurosci. **27**, 4326–4333 (2007)
2. Radovits, T., et al.: The peroxynitrite decomposition catalyst FP15 improves ageing-associated cardiac and vascular dysfunction. Mech. Ageing Dev. **128**, 173–181 (2007)
3. Wu, A.S., et al.: Iron porphyrin treatment extends survival in a transgenic animal model of amyotrophic lateral sclerosis. J. Neurochem. **85**, 142–150 (2003)
4. Szabo, C., Ischiropoulos, H., Radi, R.: Peroxynitrite: biochemistry, pathophysiology and development of therapeutics. Nat. Rev. Drug Discov. **6**, 662–680 (2007)
5. Kos, I., Benov, L., Spasojevic, I., Reboucas, J.S., Batinic-Haberle, I.: High lipophilicity of meta Mn(III) N-alkylpyridylporphyrin-based superoxide dismutase mimics compensates for their lower antioxidant potency and makes them as effective as ortho analogues in protecting superoxide dismutase-deficient Escherichia coli. J. Med. Chem. **52**, 7868–7872 (2009)
6. Haber, A., Agadjanian, H., Medina-Kauwe, L.K., Gross, Z.: Corroles that bind with high affinity to both apo and holo transferrin. J. Inorg. Biochem. **102**, 446–457 (2008)
7. Mahammed, A., Gray, H.B., Weaver, J.J., Sorasaenee, K., Gross, Z.: Amphiphilic corroles bind tightly to human serum albumin. Bioconjug. Chem. **15**, 738–746 (2004)
8. De Smidt, P.C., Versluis, A.J., Van Berkel, T.J.C.: Properties of incorporation, redistribution, and integrity of porphyrin-low-density lipoprotein complexes. Biochemistry (Mosc.) **32**, 2916–2922 (1993)
9. Rice-Evans, C., Bruckdorfer, K.R. (eds.): Oxidative stress, Lipoproteins and Cardiovascular Dysfunction. Portland Press, London (1995)
10. Kontush, A., Chapman, M.J.: Functionally defective high-density lipoprotein: a new therapeutic target at the crossroads of Dyslipidemia, inflammation, and atherosclerosis. Pharmacol. Rev. **58**, 342–374 (2006)
11. Zannis, V., Chroni, A., Krieger, M.: Role of apoA-I, ABCA1, LCAT, and SR-BI in the biogenesis of HDL. J. Mol. Med. **84**, 276–294 (2006)
12. Cheung, M.C., Albers, J.J.: Distribution of high density lipoprotein particles with different apoprotein composition: particles with A-I and A-II and particles with A-I but no A-II. J. Lipid Res. **23**, 747–753 (1982)
13. Bergmeier, C., Siekmeier, R., Gross, W.: Distribution spectrum of paraoxonase activity in HDL fractions. Clin. Chem. **50**, 2309–2315 (2004)
14. Gaidukov, L., Tawfik, D.S.: High affinity, stability, and lactonase activity of serum paraoxonase PON1 anchored on HDL with ApoA-I. Biochemistry (Mosc.) **44**, 11843–11854 (2005)
15. Aviram, M., Vaya, J.: Markers for low-density lipoprotein oxidation. Methods Enzymol. **335**, 244–256 (2001)
16. Aviram, M.: Oxidative modification of low density lipoprotein and its relation to atherosclerosis. Isr. J. Med. Sci. **31**, 241–249 (1995)
17. Aviram, M.: Interaction of oxidized low density lipoprotein with macrophages in atherosclerosis, and the antiatherogenicity of antioxidants. Eur. J. Clin. Chem. Clin. Biochem. **34**, 599–608 (1996)

# References

18. Aviram, M.: Macrophage foam cell formation during early atherogenesis is determined by the balance between pro-oxidants and anti-oxidants in arterial cells and blood lipoproteins. Antioxid. Redox Signal. **1**, 585–594 (1999)
19. Smith, J.D.: Dysfunctional HDL as a diagnostic and therapeutic target. Arterioscler. Thromb. Vasc. Biol. **30**, 151–155 (2010)
20. Leeuwenburgh, C., et al.: Mass spectrometric quantification of markers for protein oxidation by tyrosyl radical, copper, and hydroxyl radical in low density lipoprotein isolated from human atherosclerotic plaques. J. Biol. Chem. **272**, 3520–3526 (1997)
21. Perugini, C., et al.: Different mechanisms are progressively recruited to promote Cu(II) reduction by isolated human low-density lipoprotein undergoing oxidation. Free Radical Biol. Med. **25**, 519–528 (1998)
22. Urbanski, N.K., Beresewicz, A.: Generation of *OH initiated by interaction of Fe2+ and Cu+ with dioxygen; comparison with the Fenton chemistry. Acta Biochim. Pol. **47**, 951–962 (2000)
23. Mahammed, A., Gross, Z.: Highly efficient catalase activity of metallocorroles. Chem. Commun. **46**, 7040–7042 (2010)
24. Eckshtain, M., et al.: Superoxide dismutase activity of corrole metal complexes. Dalton Trans. (38), 7879–7882 (2009)
25. Esterbauer, H., et al.: Vitamin E and other lipophilic antioxidants protect LDL against oxidation. Fett Wissenschaft Technologie/Fat Sci. Technol. **91**, 316–324 (1989)
26. Bonnefont-Rousselot, D., et al.: High density lipoproteins (HDL) and the oxidative hypothesis of atherosclerosis. Clin. Chem. Lab. Med. **37**, 939 (1999)
27. Mahammed, A., Gross, Z.: Iron and manganese corroles are potent catalysts for the decomposition of peroxynitrite. Angew. Chem. Int. Ed. **45**, 6544–6547 (2006)
28. Ducrocq, C., Blanchard, B.: Peroxynitrite: an endogenous oxidizing and nitrating agent. Cell. Mol. Life Sci. **55**, 1068–1077 (1999)
29. Martin-Romero, F.J., Gutiérrez-Martin, Y., Henao, F., Gutiérrez-Merino, C.: fluorescence measurements of steady state peroxynitrite production upon SIN-1 decomposition: NADH versus dihydrodichlorofluorescein and dihydrorhodamine 123. J. Fluoresc. **14**, 17–23 (2004)
30. Aviram, M., Rosenblat, M.: Oxidative stress in cardiovascular disease: role of oxidized lipoproteins in macrophage foam cell formation and atherosclerosis. In: Fuchs, J., Podda, M., Packer, L. (eds.) Redox Genome Interactions in Health and Disease, pp. 557–590. Marcel Dekker, NY (2004)
31. Rosenblat, M., et al.: The catalytic histidine dyad of high density lipoprotein-associated serum paraoxonase-1 (PON1) is essential for PON1-mediated inhibition of low density lipoprotein oxidation and stimulation of macrophage cholesterol efflux. J. Biol. Chem. **281**, 7657–7665 (2006)
32. Bloodsworth, A., et al.: Manganese-porphyrin reactions with lipids and lipoproteins. Free Radical Biol. Med. **28**, 1017–1029 (2000)
33. Kumar, A., Goldberg, I., Botoshansky, M., Buchman, Y., Gross, Z.: Oxygen atom transfer reactions from isolated (Oxo)manganese(V) corroles to sulfides. J. Am. Chem. Soc. **132**, 15233–15245 (2010)
34. Agadjanian, H., et al.: Tumor detection and elimination by a targeted gallium corrole. Proc. Natl. Acad. Sci. U. S. A. **106**, 6105–6110 (2009)
35. Kasugai, N., et al.: Selective cell death by water-soluble Fe-porphyrins with superoxide dismutase (SOD) activity. J. Inorg. Biochem. **91**, 349–355 (2002)
36. Spasojevic, I., et al.: Pharmacokinetics of the potent redox-modulating manganese porphyrin, MnTE-2-PyP5$^+$, in plasma and major organs of B6C3F1 mice. Free Radical Biol. Med. **45**, 943–949 (2008)
37. Mandoj, F., Nardis, S., Pomarico, G., Paolesse, R.: Demetalation of corrole complexes: an old dream turning into reality. J. Porphyrins Phthalocyanines **12**, 19–26 (2008)
38. Liu, X., et al.: Development of sensitive metalloporphyrin probes for chemiluminescent imaging detection of serum proteins. Electrophoresis **30**, 3034–3040 (2009)

39. Nakano, M., et al.: A highly sensitive method for determining both Mn- and Cu–Zn superoxide dismutase activities in tissues and blood cells. Anal. Biochem. **187**, 277–280 (1990)
40. Komatsu, H., Obata, F.: An optimized method for determination of intracellular glutathione in mouse macrophage cultures by fluorimetric high-performance liquid chromatography. Biomed. Chromatogr. **17**, 345–350 (2003)
41. Parker, C.W., Fischman, C.M., Wedner, H.J.: Relationship of biosynthesis of slow reacting substance to intracellular glutathione concentrations. Proc. Natl. Acad. Sci. U. S. A. **77**, 6870–6873 (1980)
42. Haber, A., Aviram, M., Gross, Z.: Protecting the beneficial functionality of lipoproteins by 1-Fe, a corrole-based catalytic antioxidant. Chem. Sci. **2**, 295–302 (2011)
43. Ball, D.J., et al.: A comparative study of the cellular uptake and photodynamic efficacy of three novel zinc phthalocyanines of differing charge. Photochem. Photobiol. **69**, 390–396 (1999)
44. Camejo, G., et al.: Hemin binding and oxidation of lipoproteins in serum: mechanisms and effect on the interaction of LDL with human macrophages. J. Lipid Res. **39**, 755–766 (1998)
45. Gandini, S.C.M., Borissevitch, I.E., Perussi, J.R., Imasato, H., Tabak, M.: Aggregation of meso-tetrakis(4-N-methyl-pyridiniumyl) porphyrin in its free base, Fe(III) and Mn(III) forms due to the interaction with DNA in aqueous solutions: Optical absorption, fluorescence and light scattering studies. J. Lumin **78**, 53–61 (1998)
46. Sari, M.A., Battioni, J.P., Mansuy, D., Le Pecq, J.B.: Mode of interaction and apparent binding constants of meso-tetraaryl porphyrins bearing between one and four positive charges with DNA. Biochem. Biophys. Res. Commun. **141**, 643–649 (1986)
47. Jensen, M.P., Riley, D.P.: Peroxynitrite decomposition activity of iron porphyrin complexes. Inorg. Chem. **41**, 4788–4797 (2002)
48. Pasternack, R.F., Skowronek, W.R.: Catalysis of the disproportionation of superoxide by metalloporphyrins. J. Inorg. Biochem. **11**, 261–267 (1979)
49. Aviram, M., Fuhrman, B.: LDL oxidation by arterial wall macrophages depends on the oxidative status in the lipoprotein and in the cells: role of prooxidants vs. antioxidants. Mol. Cell. Biochem. **188**, 149–159 (1998)
50. Gieseg, S.P., et al.: Macrophage antioxidant protection within atherosclerotic plaques. Front Biosci. **14**, 1230–1246 (2009)
51. Szabó, C., Day, B.J., Salzman, A.L.: Evaluation of the relative contribution of nitric oxide and peroxynitrite to the suppression of mitochondrial respiration in immunostimulated macrophages using a manganese mesoporphyrin superoxide dismutase mimetic and peroxynitrite scavenger. FEBS Lett. **381**, 82–86 (1996)
52. Assreuy, J., et al.: Production of nitric oxide and superoxide by activated macrophages and killing of Leishmania major. Eur. J. Immunol. **24**, 672–676 (1994)
53. Eu, J.P., Liu, L., Zeng, M., Stamler, J.S.: An apoptotic model for nitrosative stress. Biochemistry (Mosc.) **39**, 1040–1047 (2000)
54. Pasternack, R.F., Halliwell, B.: Superoxide dismutase activities of an iron porphyrin and other iron complexes. J. Am. Chem. Soc. **101**, 1026–1031 (1979)
55. Batinić-Haberle, I., Rebouças, J.S., Spasojević, I.: Superoxide dismutase mimics: chemistry, pharmacology, and therapeutic potential. Antioxid. Redox Signal. **13**, 877–918 (2010)
56. Schwalbe, M., Dogutan, D.K., Stoian, S.A., Teets, T.S., Nocera, D.G.: Xanthene-modified and hangman iron corroles. Inorg. Chem. **50**, 1368–1377 (2011)
57. Gardner, P.R., Nguyen, D.-D.H., White, C.W.: Superoxide scavenging by Mn(II/III) tetrakis (1-methyl-4-pyridyl) porphyrin in mammalian cells. Arch. Biochem. Biophys. **325**, 20–28 (1996)
58. Dwyer, B.E., Lu, S.-Y., Laitinen, J.T., Nishimura, R.N.: Protective properties of tin- and manganese-centered porphyrins against hydrogen peroxide-mediated injury in rat astroglial cells. J. Neurochem. **71**, 2497–2504 (1998)

## References

59. Klassen, S., Rabkin, S.: The metalloporphyrin FeTPPS but not by cyclosporin A antagonizes the interaction of peroxynitrate and hydrogen peroxide on cardiomyocyte cell death. Naunyn-Schmiedeberg's Arch. Pharmacol. **379**, 149–161 (2009)
60. Asayama, S., Kasugai, N., Kubota, S., Nagaoka, S., Kawakami, H.: Superoxide dismutase as a target enzyme for Fe-porphyrin-induced cell death. J. Inorg. Biochem. **101**, 261–266 (2007)
61. Jensen, M.P., Riley, D.P.: Peroxynitrite decomposition activity of iron porphyrin complexes. Inorg. Biochem. **41**, 4788–4797 (2002)
62. Lee, J., Hunt, J.A., Groves, J.T.: Manganese porphyrins as redox-coupled peroxynitrite reductases. J. Am. Chem. Soc. **120**, 6053–6061 (1998)
63. Salvemini, D., Wang, Z.-Q., Stern, M.K., Currie, M.G., Misko, T.P.: Peroxynitrite decomposition catalysts: therapeutics for peroxynitrite-mediated pathology. Proc. Natl. Acad. Sci. U. S. A. **95**, 2659–2663 (1998)
64. Choi, I.Y., et al.: Protection by a manganese porphyrin of endogenous peroxynitrite-induced death of glial cells via inhibition of mitochondrial transmembrane potential decrease. Glia **31**, 155–164 (2000)
65. Zingarelli, B., Day, B.J., Crapo, J.D., Salzman, A.L., Szabó, C.: The potential role of peroxynitrite in the vascular contractile and cellular energetic failure in endotoxic shock. Br. J. Pharmacol. **120**, 259–267 (1997)
66. Trackey, J.L., Uliasz, T.F., Hewett, S.J.: SIN-1-induced cytotoxicity in mixed cortical cell culture: peroxynitrite-dependent and -independent induction of excitotoxic cell death. J. Neurochem. **79**, 445–455 (2001)
67. Misko, T.P., et al.: Characterization of the cytoprotective action of peroxynitrite decomposition catalysts. J. Biol. Chem. **273**, 15646–15653 (1998)
68. Rosanoff, A., Seelig, M.S.: Comparison of mechanism and functional effects of magnesium and statin pharmaceuticals. J. Am. Coll. Nutr. **23**, 501S–505S (2004)
69. Keidar, S., Aviram, M., Maor, I., Oikinne, J., Brook, J.G.: Pravastatin inhibits cellular cholesterol synthesis and increases low density lipoprotein receptor activity in macrophages: in vitro and in vivo studies. Br. J. Clin. Pharmacol. **38**, 513–519 (1994)
70. Fuhrman, B., Elis, A., Aviram, M.: Hypocholesterolemic effect of lycopene and [beta]-carotene is related to suppression of cholesterol synthesis and augmentation of LDL receptor activity in macrophages. Biochem. Biophys. Res. Comm. **233**, 658–662 (1997)
71. Fernandez, G., Spatz, E.S., Jablecki, C., Phillips, P.S.: Statin myopathy: a common dilemma not reflected in clinical trials. Cleve. Clin. J. Med. **78**, 393–403 (2011)
72. FDA Drug Safety Communication: New restrictions, contraindications, and dose limitations for Zocor (simvastatin) to reduce the risk of muscle injury. http://www.fda.gov/Drugs/DrugSafety/ucm256581.htm (2011)
73. Preiss, D., et al.: Risk of incident diabetes with intensive-dose compared with moderate-dose statin therapy. J. Am. Med. Assoc. **305**, 2556–2564 (2011)
74. Waters, D.D., et al.: Predictors of new-onset diabetes in patients treated with atorvastatin: results from 3 large randomized clinical trials. J. Am. Coll. Cardiol. **57**, 1535–1545 (2011)
75. Cheung, M.C., Zhao, X.-Q., Chait, A., Albers, J.J., Brown, B.G.: Antioxidant supplements block the response of HDL to simvastatin-niacin therapy in patients with coronary artery disease and low HDL. Artertio. Thromb. Vasc. Biol. **21**, 1320–1326 (2001)
76. Tousoulis, D., et al.: Effects of combined administration of low dose atorvastatin and vitamin E on inflammatory markers and endothelial function in patients with heart failure. Eur. J. Heart Fail. **7**, 1126–1132 (2005)
77. Coleman, R., Hayek, T., Keidar, S., Aviram, M.: A mouse model for human atherosclerosis: Long-term histopathological study of lesion development in the aortic arch of apolipoprotein E-deficient (E0) mice. Acta Histochem. **108**, 415–424 (2006)
78. Haber, A., et al.: Amphiphilic/bipolar metallocorroles that catalyze the decomposition of reactive oxygen and nitrogen species, rescue lipoproteins from oxidative damage, and attenuate atherosclerosis in mice. Angew. Chem. Int. Ed. **47**, 7896–7900 (2008)

79. Aviram, M., et al.: Pomegranate juice consumption reduces oxidative stress, atherogenic modifications to LDL, and platelet aggregation: studies in humans and in atherosclerotic apolipoprotein E-deficient mice. Am. J. Clin. Nutr. **71**, 1062–1076 (2000)
80. Kaplan, M., et al.: Pomegranate juice supplementation to atherosclerotic mice reduces macrophage lipid peroxidation, cellular cholesterol accumulation and development of atherosclerosis. J. Nutr. **131**, 2082–2089 (2001)

# Chapter 5
# Conclusions

This thesis uncovered three major phenomena characteristic of amphiphilic corroles that are of prime importance for their utilization as potential drugs for prevention or treatment of CVD.

The effects on lipoproteins (Fig. 5.1):

1. The bis-sulfonated corroles bind with high affinity to lipoproteins, with **1-Fe** displaying high selectivity for HDL2 binding. Binding of **1-Fe** is through interactions of the metal ion with two histidine moieties from yet to be determined HDL-associated protein(s), while **1-Mn** and **1-Ga** associate with the lipids of the lipoproteins due to the amphipolar structure of the corrole.
2. Conjugation of **1-Fe** to HDL and LDL efficiently protects the lipoproteins against oxidative damage and consequential function impairment, with **1-Fe**/HDL conjugates displaying improved anti-atherogenic properties relative to native HDL. Moreover, **1-Fe** reduces the pro-oxidative capacity of already oxidized LDL. On the other hand, **1-Mn** serves as a pro-oxidant under identical in vitro conditions.
3. The extremely strong binding of **1-Fe** to HDL and LDL suggests that this catalytic ROS/RNS decomposition catalyst will be carried all the way to the arterial wall, where the need for its protective action against the atherogenicity-inducing damage to oxidized lipoproteins is most crucial.
4. The highly dissimilar serum distribution of the bis-sulfonated corroles and the analogous porphyrin (does not bind to any lipoprotein) may indicate a very *different biological effect* of the compounds despite of *similar catalytic activities*.

The effects on macrophages:

1. The saturation cellular concentration of **1-Fe** (as determined by a new detection method disclosed in this research) is significant, as it is higher than the concentration of the very important cellular antioxidant enzyme SOD. On the other hand, the analogous porphyrin accumulates to very low concentrations.

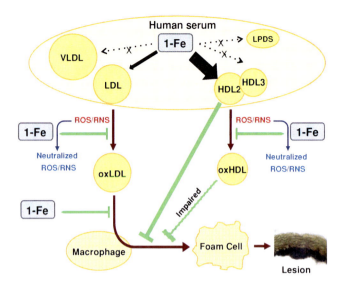

**Fig. 5.1** Summary of the effect of **1-Fe** on lipoproteins. **1-Fe** binds to HDL and LDL, protects them against modifications by ROS/RNS and saves them from oxidative-stress induced function impairment

2. Cellular uptake of corroles is mediated by serum components via a specific mechanism, without causing breakdown of the corrole. Additional research is needed to further characterize the uptake mechanism.
3. **1-Fe** and **1-Mn** are non-toxic to macrophage cells, even at high concentrations, while **1-Ga** is highly toxic. The intracellular fraction of either **1-Fe** or **1-Mn** protects the cells against primary oxidants and even from toxins that induce expression of ROS/RNS by the cells. In contrast, the intracellular fraction of the analogous porphyrin is inefficient for cellular protection against oxidative damage.
4. **1-Fe** inhibits cholesterol biosynthesis in cultured macrophages in a dose dependant manner that correlates with its intracellular levels. Co-treatment of macrophages with statins and **1-Fe** results in a combined effect, higher than the effect of each component alone. The mechanism by which **1-Fe** inhibits cellular cholesterol biosynthesis remains to be elucidated.
5. Macrophages harvested from mice treated with **1-Fe** show reduced accumulation of cholesterol and decreased cholesterol biosynthesis relative to macrophages from control mice.

The effects on atherosclerotic mice:

1. **1-Fe**-treated mice display lower atherosclerosis lesion formation relative to control mice, both in young asymptomatic mice and in an experimental design where early signs of the disease are evident before starting the treatment, a more realistic profile for a potential patient.

# 5 Conclusions

2. **1-Mn** showed limited attenuation of lesion development in the young mice model, while **1-Ga** has no effect.

These pre-clinical discoveries formed the foundation for regulated investigations that are currently in progress, required for advancing these compounds from the bench to the bedside.

# Chapter 6
# Materials and Methods

## 6.1 General Methods

### 6.1.1 Materials

- Corrole complexes were prepared according to previously described procedures [1, 2].
- Porphyrin complexes were purchased from Frontier Scientific (Logan, Utah).
- SIN-1 (Sigma) was prepared in water, saved as aliquots in −70 °C and thawed immediately before use (as it is unstable at non-basic pH).

#### 6.1.1.1 Lipoprotein Preparation

Lipoproteins were separated from serum of normal healthy volunteers by sequential ultracentrifugation [3] and dialyzed against saline with EDTA (1 mM) for removing excess salt. Protein concentration of the separated fraction was determined by the lowry method [4]. Translation to molar concentrations was calculated using a protein weight of 500 kDa for LDL. For HDL an average particle weight of 360 kDa and 45 % protein content was assumed for calculations. Fraction purity was verified by HPLC as described in Sect. 8.2.2. Lipoproteins were diluted in PBS to 1 mg protein/mL and dialyzed against PBS at 4 °C to remove the EDTA prior to oxidation experiments. OxLDL was prepared by adding 40 μM $CuSO_4$ for several hours to 1 mg protein/mL LDL until the yellow solution turned white.

#### 6.1.1.2 Protein Determination (Lowry) Assay

Samples (20 μL) were incubated with 200 μL of a color reagent for 20 min at room temperature. The color reagent was prepared by mixing 100 equivalents of 2 % $Na_2CO_3$ in 0.1 M NaOH with one equivalent of 1 % $CuSO_4$ and one equivalent of 2 % Na,K tartrate. Then 20 μL of folin reagent (diluted 1:1 in water) was added, followed by additional incubation for 20 min at room temperature. Absorption at 650 nm was measured on a PowerWave$_x$ Microplate Scanning Spectrophotometer (Bio-Tek Instruments Inc.). Protein concentrations were determined relative to a BSA standard curve (ranging up to 40 mg) [4].

### 6.1.1.3 HPLC Separation of Serum

Samples were filtered through a 0.22 μm filter and injected to a LaChrom Elite HPLC system fitted with a superose 6 10/300 GL (GE healthcare) gel filtration column and a photodiode array detector. The sample was eluted with PBS (pH = 7.4) at a flow rate of 0.5 ml/min. Chromatograms at 280 and 420 nm and the full spectrum of the run were recorded [5].

## 6.1.2 Measurement of Lipoprotein Oxidation Products

### 6.1.2.1 Conjugated Dienes

Lipoprotein samples (20 μL) were diluted with 200 μL PBS, and absorption of the samples was measured at 234 nm on a PowerWave$_x$ Microplate Scanning Spectrophotometer (Bio-Tek Instruments Inc.). Results were displayed as absorbance units or as the amount of conjugated dienes per mg of LDL or HDL protein using an extinction coefficient of 29500 $M^{-1}$ $cm^{-1}$ [6].

### 6.1.2.2 Lipid Hydroperoxides

Lipoprotein samples (20 μL) were incubated with 200 μL of a color reagent for 30 min at room temperature in the dark. The color reagent was prepared by dissolving 27.2 gr/L potassium dihydrogenphosphate, 20 gr/L potassium iodide, 0.01 gr/L sodium azide, 2 gr/L igepal, 0.1 gr/L alkylbenzyldimethyl amonium chloride, and 0.012 gr/L amonium molybdate at a final volume of 1 L, and adjusted to pH = 6.2. Then the absorption of the samples was measured at 365 nm on a PowerWave$_x$ Microplate Scanning Spectrophotometer (Bio-Tek Instruments Inc.). Results were determined using an extinction coefficient of 24600 $M^{-1}$ $cm^{-1}$ and expressed per mg of LDL or HDL protein [7].

### 6.1.2.3 Aldehydes

Lipoprotein samples (120 μL diluted with 40 μL water) were incubated with 240 μL of 0.026 M thiobarbituric acid (TBA) for 20 min at 80 °C. The TBA solution was prepared by dissolving with heating 0.375 gr TBA in 2.5 mL conc.

6.1 General Methods

HCL and 15 mL trichloroacetic acid at a final volume of 100 mL. Then the samples were centrifuged at 2000 rpm for 10 min, and 200 µL of the supernatant was collected. Absorption at 532 nm was measured on a PowerWave$_x$ Microplate Scanning Spectrophotometer (Bio-Tek Instruments Inc.). Results were determined relative to a malondialdehyde standard curve (ranging up to 50 mM) and expressed per mg of LDL or HDL protein [8].

### 6.1.2.4 Tryptophan Fluorescent

Lipoprotein samples (20 µL) were diluted with 200 µL PBS, and fluorescence of the samples was measured with excitation at 280 nm and emission at 370 nm on a Spectramax M2 Microplate Reader (Molecular Devices). Results were displayed as fluorescence units.

### 6.1.2.5 Proteins Electrophoretic Analysis

Lipoprotein samples (20 µL) were reduced and denaturated, and then subjected to SDS-PAGE on a 12 % acrylamide gel according to standard protocols. The gel was stained with coomassie blue staining, and protein bands were identified relative to a molecular weight marker, as well as by LC–MS-MS of selected bands (by The Smoler Proteomics Center, Biology Faculty, Technion).

### 6.1.2.6 Western Blot Analysis of Protein Nitration

Lipoprotein samples (20 µL) were reduced and denaturated, and then subjected to SDS-PAGE on a 12 % acrylamide gel according to standard protocols. The protein bands were transferred to a nitrocellulose membrane that was then blocked with 2 % BSA overnight at 4 °C, treated with biotinylated anti-nitrotyrosine monoclonal antibody (1:200, Cayman Chemicals) in 1 % BSA for 2 h at room temperature, and then with peroxidase-streptavidine conjugate (1:10000, Jackson ImmunoResearch Laboratories Inc.) in 1 % BSA for 1 h at room temperature. The membrane was developed using an EZ-ECL kit.

### 6.1.2.7 ELISA Detection of Protein Nitration

ELISA wells were coated by 2 h incubation at 37 °C with 100 µL of 0.4 mg/mL solution of nitrated BSA (Jackson ImmunoResearch Laboratories Inc.) in carbonate buffer (pH = 9.6). The wells were further incubated with 150 µL of 2 % BSA solution for 1 h at 37 °C. Then 50 µL of non-diluted lipoprotein samples together with 50 µL of biotinylated anti-nitrotyrosine monoclonal antibody (1:10000, Cayman Chemicals) in 2 % BSA were added for 1 h incubation at

37 °C, followed by 100 μL of peroxidase-streptavidine conjugate (1:1000, Jackson ImmunoResearch Laboratories Inc.) in 1 % BSA for 30 min at 37 °C. Finally color was developed by addition of 100 μL of 3,3′,5,5′-tetramethylbenzidine (TMB) solution for 10 min at room temperature and then 100 μL of 0.5 M $H_2SO_4$. Absorbance was measured at 450 nm on a PowerWave$_x$ Microplate Scanning Spectrophotometer (Bio-Tek Instruments Inc.). Results were determined relative to a nitrated BSA standard curve (ranging up to 100 μg/mL) and the amount of nitration per mg of HDL protein calculated under the assumption that the nitrated BSA contained 3 $NO_2$ groups per protein.

## 6.1.3 Cell Cultures

Murine J774.A1 macrophages were cultured in DMEM containing 5 % FCS in a humidified incubator. Mouse peritoneal macrophages (MPM) were harvested 4 days after intra-peritoneal injection of 3 mL thioglycolate (40 g/L) to mice. The MPM cells were washed with PBS, diluted to $10^6$ cells/mL in DMEM supplemented with 5 % FCS, plated and stored in a humidified incubator until assays measurements (no more than 5 days).

## 6.1.4 Cellular Assays

### 6.1.4.1 Cell Survival (MTT) Assay

Cells ($1 \times 10^5$ cells/well in 96-well plates) were incubated with 100 μL of 0.5 mg/mL of 3-(4,5-dimethylthiazol-2-yl)-2,5-diphenyltetrazolium bromide (MTT) in RPMI without phenol red for 1 h at 37 °C. Next, 100 μL of 10 % SDS in 0.1 M HCl were added for an overnight incubation at 37 °C. Absorption at 570 nm was measured on a PowerWave$_x$ Microplate Scanning Spectrophotometer (Bio-Tek Instruments Inc.). Results were expressed as percentage relative to values from non-treated control cells.

### 6.1.4.2 Cell Oxidation (DCFH) Assay

Cells ($1 \times 10^6$ cells/well in 12-well plates) were incubated with 10 μM 2′,7′-dichlorofluorescein diacetate (DCFH-DA) in PBS for 1 h at 37 °C, and then the cells were washed, scraped into PBS and analysed on a FACSCalibur two lasers flow cytometer with excitation at 488 nm and measurement of emission at 510–540 nm [9]. Results were expressed as mean MFI.

6.1 General Methods

### 6.1.4.3 Cell Apoptosis (DiOC$_6$) Assay

Cells (1 × 10$^6$ cells/well in 12-well plates) were incubated with 50 μM 3,3′-dihexyloxacarbocyanine iodide (DiCO$_6$) in PBS for 1 h at 37 °C, and then the cells were washed, scraped into PBS and analysed on a FACSCalibur two lasers flow cytometer with excitation at 488 nm and measurement of emission at 510–540 nm [10]. Results were expressed as MFI.

### 6.1.4.4 Cellular Cholesterol Efflux

Cells (1 × 10$^6$ cells/well in 24-well plates) were incubated for 1 h in serum-free DMEM that contained $^3$H-cholesterol (2 μCi/mL) and BSA (0.2 %) and then washed and incubated in 1 mL of DMEM containing 100 μg HDL protein. After 4 h incubation at 37 °C, 500 μL of the medium was collected, the cells were washed with PBS, 1 mL of 0.1 N NaOH was added and 500 μL was collected the next day. Medium and cellular $^3$H-cholesterol were determined by liquid scintillation counting. The percentage of cholesterol efflux was calculated as the ratio of total counts per minute in the medium divided by the total counts per minute in the medium and in the cells. HDL-mediated cholesterol efflux was calculated after subtraction of the non-specific efflux obtained in cells incubated in the absence of HDL.

### 6.1.4.5 Cellular Cholesterol Uptake

Cells (1 × 10$^6$ cells/well in 12-well plates) were incubated overnight with DMEM containing 0.2 % BSA, followed by incubation with fluorescein isothiocyanate (FITC)-labelled LDL or oxLDL (approx. 25 μg protein/mL) for 3 h at 37 °C. Then the cells were washed, scraped into PBS and analysed on a FACSCalibur two lasers flow cytometer with excitation at 488 nm and measurement of emission at 510–540 nm. Results were expressed as MFI.

### 6.1.4.6 Cellular Cholesterol Biosynthesis

Cells (2 × 10$^6$ cells/well in 6-well plates) were incubated overnight with DMEM containing 0.2 % BSA, followed by incubation with $^3$H-acetate (1 μCi/mL) or $^{14}$C-mevalonolactone (1 μCi/mL) (Amersham International, Bucks, UK) in DMEM containing 0.2 % BSA for 3 h at 37 °C. Cellular lipids were then extracted in hexane:isopropanol (3:2, v:v), and the upper phase separated by thin layer chromatography (TLC) on silica gel plates with hexane:ether:acetic acid (80:20:1.5, v:v:v). Unesterified cholesterol spots were visualized by iodine vapor (using appropriate standard), scraped into scintillation vials and counted in a $\beta$-counter [11]. 0.1 M NaOH were added to the remains of the cells, and cellular proteins were measured the following day by the Lowry method [4]. Results were expressed per mg of cell protein.

### 6.1.4.7 Cellular Cholesterol Content

Cells ($2 \times 10^6$ cells/well in 6-well plates) were treated with hexane:isopropanol (3:2, v:v) for extraction of cellular lipids [12], and the upper phase was then analyzed for cholesterol content by commercially available kits (Roche Diagnostics). Results were expressed per mg of cell protein.

### 6.1.4.8 Chemiluminescent Detection of Intracellular Iron Chelates

Cells ($1 \times 10^5$ cells/well in 96-well plates) were suspended in distilled water and frozen until measurement. Emission at 430 nm was followed in a kinetic detection mode on a Spectramax M2 Microplate Reader (Molecular Devices) after the addition of 1 mM luminol and 10 mM $H_2O_2$ (pH = 13) to the ruptured cells. The plots were then integrated from $t = 0$ to $t = 12$ and intracellular concentration determined relative to a standard curve.

## 6.1.5 *Histopathological Development of Atherosclerotic Lesions in Mice*

Heart and entire aorta were rapidly dissected out from each mouse and immersion-fixed in 3 % glutaraldehyde in 0.1 M sodium cacodylate buffer with 0.01 % calcium chloride, pH 7.4, at room temperature. The first 4 mm of the aortic arch was stained with osmium tetroxide, which colors all the lipid components a dark brown–black color thus enabling delineation of the lesion with greater accuracy. The blocks were embedded in epon resin and thin transverse sections were cut to allow greater resolution of the lesion details. The area covered by the lesion was determined by image analysis [13].

## 6.2 Experimental Procedures

### 6.2.1 *Corrole-Lipoproteins Binding*

10 μM of **1-Fe**, **1-Mn** or **1-Ga** were supplemented with 0.1 mg protein/mL of LDL (0.2 μM) or HDL (0.6 μM) and the absorbance spectrum measured. Then the solutions were extensively dialyzed to remove non-bound (and loosely bound) corrole, and the absorbance measured again. The reduction in absorbance allowed for calculating corrole concentration after dialysis, and thus for calculating the corrole/lipoprotein ratio.

## 6.2 Experimental Procedures

### 6.2.2 Corrole Distribution in Serum

#### 6.2.2.1 KBr-Density Gradient Distribution

40 μM of **1-Fe**, **1-Mn**, **1-Ga** or **2-Fe** were added to 4 mL of normal human serum and co-incubated for at least 30 min. Then the density of the solution was raised to 1.25 gr/mL by the addition of KBr and placed in the centrifugation tube. A 4 mL NaCl solution adjusted by KBr addition to a density of 1.084 gr/mL was placed above the serum, and an additional 4 mL of 1.006 gr/mL NaCl solution placed at the top. Then the tubes were treated for 48 h at 4 °C by ultracentrifugation at 40,000 rpm [3].

#### 6.2.2.2 HPLC Distribution at Low Corrole Concentrations

100 μM of **1-Fe**, **1-Mn**, **1-Ga**, **1-Al**, **1-Co**, **2-Fe** or hemin were added to normal human serum and co-incubated for at least 30 min, and then 50 μM samples were separated by HPLC as described in Sect. 8.1.4.

#### 6.2.2.3 HPLC Distribution for Serum Overloaded with Corrole

**1-Fe** was added to the human serum in 1:1 volume ratio, giving a final corrole concentration of 500 μM and two-fold diluted serum. These solutions were dialyzed against three replacements of saline-EDTA for removing all non-tightly bound corrole, and then 50 μM samples were separated by HPLC as described in Sect. 8.1.4.

#### 6.2.2.4 HPLC Distribution in Serum of Mice Treated with Corrole

Two 10-weeks old male C57Bl/6 mice were intraperitoneally injected with 200 μL solution of 1 mM **1-Fe** or **1-Ga**, and blood was collected 40 min later. After 30 min the blood was centrifuged at 3000 rpm for 10 min, the serum was collected and 50 μM samples were separated by HPLC as described in Sect. 8.1.4.

### 6.2.3 Spectral Changes in the Presence of Lipoprotein Components

#### 6.2.3.1 CD Spectra with Lipoproteins

60 μM of **1-Fe** or **1-Mn** or 15 μM of **1-Ga** were incubated for 10 min with 2.5 mg protein/mL HDL2 (∼15 μM) or 2.5 mg/mL LDL (∼5 μM), and then CD spectra at the visible range were measured on a JASCO J-810 CD spectropolarimeter

using a data pitch of 1 nm, a bandwidth of 1 nm, a detector response of 1 s and a scan speed of 100 nm/min.

### 6.2.3.2 Electronic Spectra with Amino Acids

Concentrated solutions of MeHis, MeMet, MeTyr, EtCys, MeArg and MeLys (all containing hydrochloride in their formulation) were prepared and brought to neutral pH by the addition of NaOH. Microliters of these solutions were added to a 50 μM solution of **1-Fe** or **1-Mn**, and the electronic spectrum was measured.

### 6.2.3.3 Electronic Spectra with Liposomes

Di-oleoyl-phosphatidylcholine, unesterified cholesterol and cholesterol oleate ester were each dissolved in a methamol:chloroform (1:2) solution, and mixed to a final concentration of 2.6 mM, 90 and 90 μM (in 1 ml), respectively [14]. The solvent was evaporated under $N_2$, and 100 μL of tris-buffered saline (TBS: 20 mM tris, 150 mM NaCl, pH = 8) was added. A 1 min vortex followed by 2 min sonication and another 1 min vortex were applied, followed by 30 min shaking at 37 °C. For constructing lipidic liposomes, another 900 μL of TBS were added for 2 h incubation at 37 °C under shaking. PON1 or apoE were then added to a concentration of 0.2 μM, and another 2 h shaking at 37 °C was applied. Microliters of these solutions were added to a 30 μM solution of **1-Fe** or **1-Mn**, and the electronic spectrum was measured.

### 6.2.3.4 Electronic Spectra in the Serum of Knockout Mice

Blood was collected from C57Bl/6 mice and from apoE, PON1, and apoAI deficient mice (on a background of C57Bl/6). After 30 min the blood was centrifuged at 3000 rpm for 10 min, and the serum collected. **1-Fe** was added to the mice serum to a final concentration of 100 μM, and 30 μM samples were separated by HPLC as described in Sect. 8.1.4.

### 6.2.3.5 Electronic Spectra with PON1-Treated HDL2

500 μg protein/mL HDL2 (3 μM of HDL2, 12 μM of apoAI) were incubated while shaking for 2 h at 37 °C with or without 12 μM of PON1. To these solutions 12 μM of **1-Fe** was added, and the absorbance spectrum measured.

6.2 Experimental Procedures

## *6.2.4 Lipoprotein Oxidation*

### 6.2.4.1 Dose-Dependent Effect of Corroles on Copper-Ions-Induced Oxidation

LDL or HDL (0.1 mg protein/mL) solutions in PBS were incubated for 30 min at room temperature without any additive or with **1-Mn**, **1-Fe** or **1-Ga** at concentrations of 0.5, 2.5 and 5 μM. Oxidation was initiated by addition of a freshly prepared $CuSO_4$ solution to a concentration of 5 μM and the solutions were incubated at 37 °C, 2 h for LDL and 5 h for HDL. Lipoprotein oxidation was determined by measuring the amount of lipid hydroperoxides and aldehydes formation (Sect. 8.1.4, page 61).

### 6.2.4.2 Effect of Corroles on Kinetics of Copper-Ions-Induced Oxidation

LDL or HDL (0.1 mg protein/mL) solutions in PBS were incubated for 30 min at room temperature without any additive or with 2.5 μM **1-Mn** or **1-Fe**. Oxidation was initiated by addition of a freshly prepared $CuSO_4$ solution to a concentration of 5 μM and conjugated dienes formation was continuous monitored for 120 min while formation of lipid hydroperoxides and aldehydes was measured every 15 min (Sect. 8.1.4, page 61).

### 6.2.4.3 Dose-Dependent Effect of SIN-1

HDL (1 mg protein/mL) in PBS containing 100 μM DTPA was incubated overnight without or with SIN-1 at concentrations of 100, 200, 300, 500 or 1000 μM SIN-1 at 37 °C. Lipoprotein oxidation was determined by measuring the formation of conjugated dienes, lipid hydroperoxides, aldehydes, tryptophan fluorescence, protein electophoretic pattern, and protein nitration by western blot and ELISA (Sect. 8.1.4, page 61).

### 6.2.4.4 Dose-Dependent Effect of Corroles on SIN-1-Induced Oxidation

HDL (1 mg protein/mL) solutions in PBS containing 100 μM DTPA were incubated for 30 min at room temperature without any additive or with **1-Mn** or **1-Fe** at concentrations of 10, 25 and 50 μM. Oxidation was initiated by addition of SIN-1 to a concentration of 500 μM and incubation 2 h at 37 °C. Lipoprotein oxidation was determined by measuring the formation of conjugated dienes, protein electophoretic pattern, and protein nitration by western blot (Sect. 8.1.4, page 61).

### 6.2.4.5 Effect of Corroles on Kinetics of SIN-1-Induced Oxidation

HDL (1 mg protein/mL) solutions in PBS containing 100 µM DTPA were incubated for 30 min at room temperature without any additive or with 50 µM **1-Mn** or **1-Fe**. Oxidation was initiated by addition of SIN-1 to a concentration of 500 µM, and conjugated dienes formation was continuous monitored for 120 min (Sect. 8.1.4, page 61).

## 6.2.5 Lipoprotein Functions

### 6.2.5.1 Anti-Atherogenic Functions of HDL

HDL (1 mg protein/mL) in PBS was incubated with or without 50 µM of **1-Fe** for 10 min (until **1-Fe** containing solutions turned green), and then 500 µM SIN-1 or 40 µM CuSO$_4$ were added for overnight incubation at 37 °C. At the end of oxidation solutions were analyzed for the following biological activities:

- *HDL-mediated cholesterol efflux*: Cholesterol efflux from J774.A1 cells (1 × 10$^6$ cells/well in 24-well plates) was conducted as described in the general procedures (Sect. 8.1.7, page 64) by using the various HDL samples.
- *Protection against oxLDL-induced oxidation*: J774.A1 cells (1 × 10$^6$ cells/well in 12-well plates) were treated with 50 µg protein/mL of the various HDL samples in serum free media. After 30 min incubation at 37 °C, oxLDL was added at a 25 µg protein/mL concentration for 4 h, after which cells were washed and cellular oxidative stress was measured by the DCFH assay (Sect. 8.1.7, page 64). oxLDL-induced oxidative stress was calculated by subtracting the basal cellular oxidative stress (without oxLDL and without HDL) from each recorded value.
- *Protection against tunicamycin-induced apoptosis*: J774.A1 cells (1 × 10$^6$ cells/well in 12-well plates) were treated with 100 µg protein/mL of the various HDL samples in serum free media. After 30 min incubation at 37 °C, tunicamycin was added at a 1 µg/mL concentration for overnight incubation, after which cells were washed and cellular viability was measured by the DiCO$_6$ assay (Sect. 8.1.7, page 64). Tunicamycin-induced apoptosis was calculated by subtracting recorded values from the value of untreated cells (without tunicamycin and without HDL).

### 6.2.5.2 Pro-Oxidative Activity of oxLDL

oxLDL (1 mg protein/mL) was incubated overnight at 37 °C in the absence or presence of 50 µM **1-Fe**. These oxLDL solutions were added at a 25 µg protein/mL concentration for 4 h to J774A.1 cells (1 × 10$^6$ cells/well in 12-well plates) in serum free media. Then the cells were washed and cellular oxidative stress was measured using the DCFH assay (Sect. 8.1.7, page 64). oxLDL-induced oxidative stress was calculated by subtracting the basal cellular oxidative stress (without oxLDL) from the recorded values.

### 6.2.6 Interactions of Corrole and Porphyrin Complexes with Cells

#### 6.2.6.1 Detection of 1-Ga by Fluorescent Microscopy

J774.A1 cells were seeded on cover slips and incubated in DMEM containing 5 % FCS with or without 20 µM **1-Ga** at 37 °C for 30 min. Then the cells were washed, fixed with 4 % paraformaldehyde for 30 min, suspended in DAPI-containing mounting media, and examined by fluorescence microscopy using a LSM 510 Meta laser scanning confocl system (Carl Zeiss). DAPI was detected by excitation at 405 nm, and corrole by excitation at 561 nm with a 600 nm long pass emission filter.

#### 6.2.6.2 Detection of 1-Ga by Flow Cytometry

J774.A1 cells ($1 \times 10^6$ cells/well in 12-well plates) were incubated in DMEM containing 5 % FCS without or with 5 or 20 µM **1-Ga** at 37 °C for 30 min. Then the cells were washed, scraped into PBS and analysed on a FACSCalibur two lasers flow cytometer with excitation at 635 nm and emission at 653–669 nm.

#### 6.2.6.3 Standard Curve for Analysis of 1-Ga in Cellular Debris

J774.A1 cells ($1 \times 10^5$ cells/well in 96-well plates) were washed, suspended in distilled water and frozen until detection. Cells were then thawed and increasing concentrations of **1-Ga** were added. Fluorescence measurement ($\lambda_{ex} = 430$ nm, $\lambda_{em} = 615$ nm) allowed for the construction of a calibration curve.

#### 6.2.6.4 Time-Dependent Cellular Accumulation of 1-Ga

J774.A1 cells ($1 \times 10^5$ cells/well in 96-well plates) were incubated in DMEM containing 5 % FCS with or without 20 µM of **1-Ga** at 37 °C for 1, 3 or 24 h. Then the cells were washed, suspended in distilled water and frozen until detection. Intracellular corrole was detected and quantified relative to a standard curve (in the presence of cellular debris) by measurement of fluorescence ($\lambda_{ex} = 430$ nm, $\lambda_{em} = 615$ nm). Cell survival was determined in parallel by the MTT assay (Sect. 8.1.7, page 64).

### 6.2.6.5 Detection of 1-Fe and 1-Mn in Cell-Free Systems

1 mM luminol was added to solutions containing 1000 nM, 100 nM or 10 nM of **1-Fe** or **1-Mn** under various pH conditions: 0.1 M NaOH (pH = 13), 50 mM glycine buffer at pH = 9, or 50 mM phosphate buffer at pH = 7. Next, 1 mM or 10 mM of $H_2O_2$ were added, and kinetics of emission at 430 nm was immediately measured.

### 6.2.6.6 Standard Curve for Analysis of 1-Fe in Cellular Debris

J774.A1 cells ($1 \times 10^5$ cells/well in 96-well plates) were washed, suspended in distilled water and frozen until detection. Cells were then thawed and increasing concentrations of **1-Fe** were added. Next, 1 mM luminol were added at pH = 13. Finally, 10 mM $H_2O_2$ were added, and kinetics of emission at 430 nm was immediately measured. Integration of the 12 min kinetic plots allowed for the construction of a calibration curve.

### 6.2.6.7 Time-Dependent Cellular Accumulation of 1-Fe

J774.A1 cells ($1 \times 10^5$ cells/well in 96-well plates) were incubated in DMEM containing 5 % FCS with or without 20 μM of **1-Fe** at 37 °C for 1, 3 or 24 h. Then the cells were washed, suspended in distilled water and frozen until detection. Intracellular corrole was detected and quantified relative to a standard curve (in the presence of cellular debris) by chemiluminescence (Sect. 8.1.7, page 64). Cell survival was determined in parallel by the MTT assay (Sect. 8.1.7, page 64).

### 6.2.6.8 Time-Dependent Removal of 1-Fe from Cells

J774.A1 cells ($1 \times 10^5$ cells/well in 96-well plates) were incubated in DMEM containing 5 % FCS with or without 20 μM of **1-Fe** at 37 °C for 24 h, washed and further incubated in corrole-free and serum-free medium at 37 °C for 3 or 24 h. Then the cells were washed, suspended in distilled water and frozen until detection. Intracellular corrole was detected and quantified relative to a standard curve (in the presence of cellular debris) by chemiluminescence (Sect. 8.1.7, page 64). Cell survival was determined in parallel by the MTT assay (Sect. 8.1.7, page 64).

### 6.2.6.9 Dose-Dependent Cellular Accumulation of 1-Fe

J774.A1 cells ($1 \times 10^5$ cells/well in 96-well plates) were incubated in DMEM containing 5 % FCS with increasing concentrations of **1-Fe** for 2 h at 37 °C. Then the cells were washed, suspended in distilled water and frozen until detection. Intracellular corrole was detected and quantified relative to a standard curve (in the presence of cellular debris) by chemiluminescence (Sect. 8.1.7, page 64).

### 6.2.6.10 Standard Curve for Analysis of Iron Porphyrins in Cellular Debris

J774.A1 cells ($1 \times 10^5$ cells/well in 96-well plates) were washed, suspended in distilled water and frozen until detection. Cells were then thawed and increasing concentrations of **2-Fe**, **3-Fe** or hemin were added. Next, 1 mM luminol were added at pH = 13. Finally, 10 mM $H_2O_2$ were added, and kinetics of emission at 430 nm was immediately measured. Integration of the 12 min kinetic plots allowed for the construction of a calibration curve.

### 6.2.6.11 Relative Cell Accumulation of 1-Fe and Related Porphyrins

J774.A1 cells ($1 \times 10^5$ cells/well in 96-well plates) were incubated in DMEM containing 5 % FCS with or without 20 μM of **1-Fe**, **2-Fe**, **3-Fe** or hemin at 37 °C for 24 h. Then the cells were washed, suspended in distilled water and frozen until detection. Intracellular corrole/porphyrin was detected and quantified relative to a standard curve (in the presence of cellular debris) by chemiluminescence (Sect. 8.1.7, page 64). Cell survival was determined in parallel by the MTT assay (Sect. 8.1.7, page 64).

## 6.2.7 *Effect of Corrole and Porphyrin Complexes on Cells*

### 6.2.7.1 Effect of 1-Fe and 2-Fe Against Oxidative-Stress Induced Death

J774.A1 cells ($1 \times 10^5$ cells/well in 96-well plates) were incubated in DMEM containing 5 % FCS with: medium free of additives or with medium containing 20 μM of **1-Fe** or **2-Fe** at 37 °C for 24 h. Then cells were washed and treated with medium alone or with medium containing: 500 μM $H_2O_2$, 1000 μM SIN-1 or 25 μg/mL oxLDL for 24 h or with a combination of 100 ng/mL LPS and 20 ng/mL INFγ for 48 h at 37 °C. Cell survival was determined by the MTT assay (Sect. 8.1.7, page 64).

### 6.2.7.2 Effect of 1-Fe on Cellular Cholesterol Flux

J774.A1 cells ($1 \times 10^6$ cells/well in 12-well plates for uptake and 24-well plates for efflux) were incubated in DMEM containing 5 % FCS with or without 20 μM of **1-Fe** at 37 °C for 24 h. Then the cells were washed and analyzed for cholesterol uptake and efflux (Sect. 8.1.7, page 64). The lipoproteins used for the examination *were not* previously treated with corrole.

### 6.2.7.3 Dose-Dependent Effect of 1-Fe on Cholesterol Biosynthesis

J774.A1 cells ($2 \times 10^6$ cells/well in 6-well plates) were incubated for 24 h with or without 5, 20 or 50 µM of **1-Fe**. After washing the cells, cholesterol biosynthesis from acetate was determined (Sect. 8.1.7, page 64). Intracellular corrole concentration was determined in parallel following the same treatment (Sect. 8.1.7, page 64).

### 6.2.7.4 Compared and Combined Effects of 1-Fe and Statins on Cholesterol Biosynthesis

J774.A1 cells ($2 \times 10^6$ cells/well in 6-well plates) were incubated for 24 h with or without 20 µM **1-Fe**, 200 µM pravastatin (PS), 20 µM fluvastatin (FS), or a combination of corrole and statin. After washing the cells, cholesterol biosynthesis from acetate was determined (Sect. 8.1.7, page 64).

### 6.2.7.5 Effects of 1-Fe Administration to Mice on MPM Cholesterol Biosynthesis

12-weeks old male $E^0$ mice, 12-weeks old female $E^0$ mice, or 30-weeks old male $E^0$ mice (6 in each group) received tap water with no additive or containing **1-Fe** (at a dosage of 10 mg/kg/day). After 12 weeks MPM cells were harvested (Sect. 8.1.6), and analyzed for cholesterol biosynthesis and cholesterol content (Sect. 8.1.7, page 64).

### 6.2.7.6 Mechanism Behind 1-Fe Effect on Cholesterol Biosynthesis

J774.A1cells ($2 \times 10^6$ cells/well in 6-well plates for cholesterol biosynthesis and $1 \times 10^6$ cells/well in 12-well plates for the other assays) were incubated for 24 h with or without 20 µM of **1-Fe**. After washing the cells, cholesterol biosynthesis from mevalone (Sect. 8.1.7, page 64), HMGCR mRNA expression, HMGCR protein expression or HMGCR protein phosphorylation were determined.

- *HMGCR mRNA expression*: Total RNA was extracted with TRI-reagent (Molecular Research Center, Inc.). Final total RNA concentration was determined spectrophotometrically by measuring the absorbency at 260 nm. cDNA was generated from 1 µg of total RNA using M-MLV reverse transcriptase (Promega) and random primers (Promega). The reaction was carried out at 37 °C for 60 min followed by 5 min at 95 °C. The products of the reaction were subjected to PCR amplification: first step at 94 °C for 1 min, second step at 95 °C for 1 min, third step at 55 °C for 1 min, fourth step at 72 °C for 1.5 min (steps 2, 3 and 4 were cycled for 35 times), and the last step was performed at 72 °C for 8 min. The amplified transcripts were separated on a 1 % (w/vol.) agarose gel containing ethidium bromide and visualized with ultraviolet transillumination.

*Primer for the HMGCR*:
Forward: 5′-GGGACGGTGACACTTACCATCTGTATGATG-3′
Reverse: 5′-ATCATCTTGGAGAGATAAAACTGCCA-3′.
*Primers for the house keeping gene glyceraldehyde 3-phosphate dehydrogenase (GAPDH)*:
Forward: 5′-CTGCCATTTGCAGTGGCAAAGTGG-3′
Reverse: 5′-TTGTCATGGATGACCTTGGCCAGG-3′.

- *HMGCR protein expression and phosphorylation:* HMGCR protein expression and phosphorylation and protein expression of the house keeping gene $\beta$-actin were detected by western blot analysis using cell lysates containing 20 μg protein on a 10 % SDS-PAGE. The membrane was blocked with 2 % BSA in TBST for 1.5 h, followed by treatment with a primary antibody (in TBST containing 1 % BSA) for 2 h and a secondary antibody (in TBST containing 1 % BSA) for 1 h, all at room temperature. The membrane was then developed using an EZ-ECL kit. The antibodies used were as followed:

| Protein | Primary antibody | Secondary antibody |
|---|---|---|
| HMGCR | Goat anti HMGCR (1:300) Santa cruz | Rabbit anti goat (1:5000) Jackson Laboratories |
| Phospho-HMGCR | Rabbit anti Phospho-HMGCR (1:1000) Upstate | Goat anti rabbit (1:5000) Jackson Laboratories |
| $\beta$-actin | Mouse anti $\beta$-actin (1:5000) Sigma | Rabbit anti goat (1:10000) Jackson Laboratories |

- *HMGCR activity in cell free systems:* The activity of HMGCR without an inhibitor, in the presence of pravastatin or following addition of 0.1 μM of **1-Fe** was examined by the HMG-CoA reductase activity kit (Sigma).

## *6.2.8 The Effect of Corroles on the Development of Atherosclerosis*

### 6.2.8.1 Effect of Corrole Treatment on Young Mice

At an age of 12 weeks, 24 male $E^0$ mice were divided randomly to 4 groups of 6 mice each. The groups differed only in the type of drinking water: no additive, and water containing 0.04 mM of either **1-Mn**, **1-Fe** or **1-Ga**. Fluid consumption by the groups receiving **1-Mn** and **1-Ga** was ~5 mL/mouse/day, which equals to 0.2 mg per mouse per day (or 10 mg/kg/day). The group receiving **1-Fe** was found to drink somewhat larger amounts (~6 mL/mouse/day). After 10 weeks the mice were sacrificed and blood samples, heart with attached aorta and MPM were collected from all mice.

### 6.2.8.2 Effect of Corrole Treatment on Aged Mice

At an age of 30 weeks, 10 male $E^0$ mice were divided randomly to 2 groups of 5 mice each. Treated mice received 10 mg/kg/day of **1-Fe** within their drinking water. After 12 weeks the mice were sacrificed and blood samples, heart with attached aorta and MPM were collected from all mice.

## References

1. Mahammed, A., Gross, Z.: Iron and manganese corroles are potent catalysts for the decomposition of peroxynitrite. Angew. Chem. Int. Ed. **45**, 6544–6547 (2006)
2. Saltsman, I., et al.: Selective substitution of corroles: nitration, hydroformylation, and chlorosulfonation. J. Am. Chem. Soc. **124**, 7411–7420 (2002)
3. Aviram, M.: Plasma lipoprotein separation by discontinuous density gradient ultracentrifugation in hyperlipoproteinemic patients. Biochem. Med. **30**, 111–118 (1983)
4. Lowry, O.H., Rosebrough, N.J., Farr, A.L., Randall, R.J.: Protein measurement with the Folin phenol reagent. J. Biol. Chem. **193**, 265–275 (1951)
5. Mira, R., Rachel, K., Michael, A.: Atherosclerosis **187**, 74.e1–74.e10 (2006)
6. Aviram, M., Vaya, J.: Methods Enzymol. **335**, 244–256 (2001)
7. El-Saadani, M., Esterbauer, H., El-Sayed, M., Goher, M., Nassar, A.Y., Jurgens, G.: A spectrophotometric assay for lipid peroxides in serum lipoproteins using a commercially available reagent. J. Lipid. Res. **30**, 627–630 (1986)
8. Buege, J.A., Aust, S.D.: Microsomal lipid peroxidation. Methods Enzymol. **52**, 302–310 (1978)
9. LeBel, C.P., Ischiropoulos, H., Bondy, S.C.: Evaluation of the probe 2′,7′-dichlorofluorescin as an indicator of reactive oxygen species formation and oxidative stress. Chem. Res. Toxicol. **5**, 227–231 (2002)
10. Fuhrman, B., Gantman, A., Aviram, M.: Paraoxonase 1 (PON1) deficiency in mice is associated with reduced expression of macrophage SR-BI and consequently the loss of HDL cytoprotection against apoptosis. Atherosclerosis **211**, 61–68 (2010)
11. Oram, J.F., Albers, J.J., Bierman, E.L.: Rapid regulation of the activity of the low density lipoprotein receptor of cultured human fibroblasts. J. Biol. Chem. **255**, 475–485 (1980)
12. Hara, A., Radin, N.S.: Lipid extraction of tissues with a low toxicity solvent. Anal. Biochem. **90**, 420–426 (1978)
13. Coleman, R., Hayek, T., Keidar, S., Aviram, M.: A mouse model for human atherosclerosis: long-term histopathological study of lesion development in the aortic arch of apolipoprotein E-deficient (E0) mice. Acta Histochem. **108**, 415–424 (2006)
14. Efrat, M., Rosenblat, M., Mahmood, S., Vaya, J., Aviram, M.: Di-oleoyl phosphatidylcholine (PC-18:1) stimulates paraoxonase 1 (PON1) enzymatic and biological activities: In vitro and in vivo studies. Atherosclerosis **202**, 461–469 (2009)

# Appendix
# 1-Fe Catalyses DCFH Oxidation to Its Fluorescent Derivative

Measurements of cellular oxidative stress are achieved by using DCFH, a non-fluorescent molecule that becomes fluorescent following its oxidation by intracellular peroxides to DCF. Cells are then examined by flow cytometry, with an excitation wavelength of 485 nm and emission detection at 515–545 nm. Because **1-Fe** contains absorbance bands throughout the visible range, it may cause quenching of DCF fluorescence, and thus interfere with the measurement, incorrectly indicating reduced cellular oxidativity. To examine this possibility, DCFH was incubated with tert-butyl hydroperoxide in a cell free system, followed by addition of **1-Fe**, and fluorescence measurement. Not only that the fluorescence was not reduced, it was actually dramatically enhanced due to the addition of the corrole, indicating that **1-Fe** apparently catalyses the oxidation of DCFH by peroxides. Two important conclusions may be drawn from here. First, the reduced cellular oxidative stress displayed by **1-Fe**/HDL conjugates relative to non-conjugated HDL (Fig. 3.25, p. 33) was not an artifact. Second, the increased DCFH signal for cells containing intracellular corrole may serve as additional evidence for the presence of the corrole within the cells rather than signal increased cellular oxidative stress (Sect. 3.4, p. 41).